MICROSCOPY HANDBOOKS 16

Light-Element Analysis in the Transmission Electron Microscope: WEDX and EELS

P.M. Budd and Peter J. Goodhew

MicroStructural Studies Unit
University of Surrey
Guildford, UK

Oxford University Press · Royal Microscopical Society · 1988

Oxford University Press, Walton Street, Oxford OX2 6DP
Oxford New York Toronto
Delhi Bombay Calcutta Madras Karachi
Petaling Jaya Singapore Hong Kong Tokyo
Nairobi Dar es Salaam Cape Town
Melbourne Auckland
and associated companies in
Beirut Berlin Ibadan Nicosia

Royal Microscopical Society
37/38 St. Clements
Oxford OX4 1AJ

Oxford is a trade mark of Oxford University Press

Published in the United States
by Oxford University Press, New York

© *Royal Microscopical Society 1988*

All rights reserved. No part of this publication may be reproduced, stored in a retrieval system, or transmitted, in any form or by any means, electronic, mechanical, photocopying, recording, or otherwise, without the prior permission of Oxford University Press

This book is sold subject to the condition that it shall not, by way of trade or otherwise, be lent, re-sold, hired out, or otherwise circulated without the publisher's prior consent in any form of binding or cover other than that in which it is published and without a similar condition including this condition being imposed on the subsequent purchaser

British Library Cataloguing in Publication Data
Budd, P.M.
Light-element analysis in the transmission
electron microscope : WEDX and EELS. —
(Microscopy handbooks; 16).
1. Electron microscopy   2. Materials —
Testing
I. Title   II. Goodhew, Peter J.   III. Series
620.1'1299   TA417.23
ISBN 0-19-856417-1

Library of Congress Cataloging in Publication Data
Budd, Pamela M. (Pamela M.)
Light-element analysis in the transmission
electron microscope, WEDX and EELS.
(Microscopy handbooks ; 16)
1. Electron energy loss spectroscopy. 2. Windowless
energy-dispersive X-ray analysis. 3. Electron microscope,
Transmission. 4. Light element — Analysis.   I. Goodhew,
Peter J.   II. Title.   III. Series.
QD96.E44B83   1988   543'.0812   87-31234
ISBN 0-19-856417-1

Set by Grestun Graphics
Printed in Great Britain by Express Litho Service (Oxford)

Preface

There are other handbooks in the RMS series which deal with analysis in the electron microscope and it would be reasonable to ask 'why another?' The answer is quite straightforward: The determination of light elements poses particular problems, and exploits techniques, not normally encountered in conventional X-ray analysis. This handbook concentrates on two techniques which are rapidly gaining currency in transmission electron microscopy (TEM). These are windowless energy-dispersive X-ray analysis, which we will call WEDX, and electron energy loss spectrometry, which is widely known as EELS. Other handbooks in the series provide an excellent introduction to X-ray microanalysis as practised both by materials scientists and by biological and medical scientists. In this handbook we explain the ways in which microscopists from any discipline can extend their analytical capability to elements as light as helium or even hydrogen. We will make no attempt to address the problems of analysing solid specimens, which has been treated by Morgan (1985) (see References) and will confine our discussion to specimens thin enough to be imaged in a TEM.

University of Surrey P.M.B,
September 1987 P.J.G.

Acknowledgements

Experience with light element analysis does not come easily. We have been particularly fortunate to have had access to up-to-date hardware and computing facilities and the cooperation of keenly interested microscopists at Surrey and elsewhere. Among those whose help we are delighted to acknowledge are Peter Statham and Steve Vale of Link Analytical, Peter Bovey and Ian Wardell of VG Scientific, and Dawn Chescoe and Vernon Power of our own MicroStructural Studies Unit. Support for the hardware has come from the University of Surrey, SERC, LINK, and VG. Some of the analytical examples have been provided by our colleagues Ray Cox, Robin Payne, and Nick Gregg.

Contents

1 Introduction ... 1

2 Windowless energy-dispersive X-ray analysis ... 5
   2.1 The detector ... 8
   2.2 The electronics ... 11
   2.3 Escape and sum peaks ... 12
   2.4 Performing an analysis ... 14
   2.5 Qualitative interpretation of the spectrum ... 16
   2.6 Mapping ... 16

3 WEDX: Quantification and limits ... 18
   3.1 Determination of k factors ... 20
   3.2 Absorption of X-rays in the specimen ... 22
   3.3 Performing a quantitative analysis ... 26
   3.4 Methods of determining local specimen thickness ... 27
   3.5 Accuracy and detection limits ... 30

4 Electron energy-loss spectrometry ... 35
   4.1 Description of the spectrometer ... 35
   4.2 Spectrometric parameters ... 38
   4.3 Coupling of the spectrometer to the microscope ... 39
   4.4 Microscope and sample parameters ... 42
   4.5 Calibration of the spectrum ... 42
   4.6 Qualitative interpretation of an EEL spectrum ... 43

5 Quantification of EELS data ... 48
   5.1 Quantification procedure ... 48
   5.2 Detection limits and accuracy ... 53
   5.3 Limitations of the technique ... 53
   5.4 Other information obtainable ... 55

6 Comparison of the two techniques ... 59
   6.1 Range of elements detectable ... 59
   6.2 Thickness limitations ... 61
   6.3 Overlap problems ... 62
   6.4 Contamination ... 63
   6.5 Other points of comparison ... 66

Appendix 1. Some of the more useful low-energy X-ray lines   67
Appendix 2. Mass absorption coefficients for light-element K lines   68
Appendix 3. Some of the low-energy edges   69

References   70

Index   72

Introduction

The usual definition of a 'light element' in the context of electron microscopy (EM) analysis arises from the difficulty of detecting, in an energy-dispersive detector with a conventional beryllium window, characteristic X-rays emitted by elements of atomic number smaller than sodium. Conventionally the ten elements hydrogen to neon are therefore referred to as the 'light elements'. Since many of the elements commonly encountered in nature and engineering are among this group there is widespread interest in being able to determine them with high spatial resolution in the transmission electron microscope (TEM).

There are, in principle, three fairly direct ways of detecting light elements in a TEM. A small fraction of the high-energy electrons in the electron beam will excite one or more atoms during their passage through the specimen. This excitation will often consist of the knocking-out of an inner-shell electron, leaving an energetic atom with a vacant electron state. This atom must later relax to its ground state and in doing so must emit a discrete amount of energy. Two of the common ways for this to happen are by the emission of a characteristic X-ray or by the ejection of an outer electron carrying a characteristic energy, a so-called 'Auger electron'. Figure 1.1 illustrates these processes.

After the high-energy electron beam has interacted with the specimen we could therefore try to measure one of the following:

1. The energy lost by each primary electron in exciting atoms of the sample. This will be characteristic of a particular electron within a particular atom and therefore provides an excellent analytical signal.
2. The energy of characteristic X-rays which result from the relaxation of some excited atoms.
3. The energy of characteristic Auger electrons which are emitted as an alternative to X-rays.

All three approaches are in widespread use and have led to the development of electron energy loss spectrometry (EELS), X-ray analysis (often called EDX, for energy-dispersive X-ray analysis) and Auger electron spectrometry (AES). However the choice of technique for the determination of light elements in the TEM is governed by two major considerations: the size of the signal which can be detected and the practical constraint of installing the necessary detector on a TEM.

We first consider the signal. The cross-section, Q, for the excitation (i.e. knocking-out) of an inner-shell electron by a primary electron of energy E_0 is of the form

$$Q = \frac{cZ \ln (E_0/E_c)}{E_0 E_c} \text{ m}^2$$

2   Light element analysis in the T.E.M.

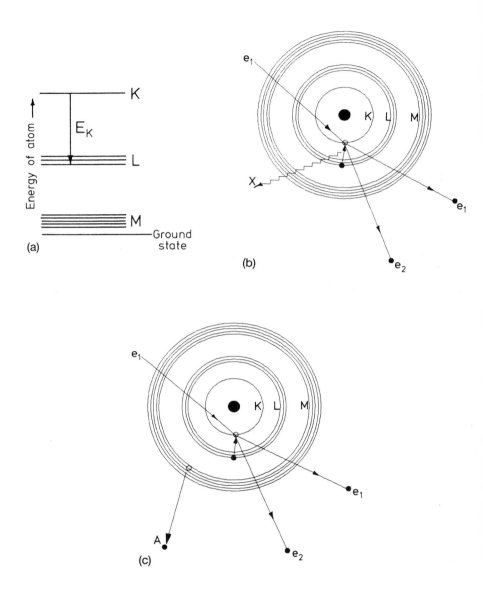

Fig. 1.1. The two relaxation processes which an excited atom can undergo: The K, L and M energy levels of an excited atom are shown in (a). When a K-shell vacancy is filled by an L electron, as shown, the energy E_K can be emitted as an X-ray or an Auger electron. In both (b) and (c) a primary electron (e_1) has knocked out an inner (K-shell) electron (e_2), which is replaced a short time later by an L_3 electron. In (b) the energy is emitted as a Kα X-ray while in (c) it is carried off as the kinetic energy of an Auger electron, in this case an M_2 electron.

where E_c is the critical ionization energy necessary to eject the electron and Z is the atomic number of the element. The constant c is different for each shell (K, L, M, etc.). Since E_c is approximately proportional to Z^2, the cross-section is larger for elements of low atomic number. This is one of the few factors which acts in favour of light-element analysis! The average number of atoms, n, excited by a single electron as it passes through a (thin) elemental specimen of thickness t is approximately

$$n = QN\rho t/A \quad \text{(excitations/electron)}$$

where N is Avogadro's number, A is the atomic mass of the element and ρ is its density. In most situations n will be very small. For example for the excitation of K-shell electrons in 10 nm of copper by 100 keV electrons, n is only 2.5×10^{-4}. Clearly, in a real analysis the concentration of the element to be analysed will be less than 100% and the number of excitations will be proportionally smaller.

The excited atoms can relax in one of two ways, yielding either a characteristic X-ray or an Auger electron. The fluorescence yield, w, (the fraction of excitations which gives an X-ray) is strongly dependent on atomic number, Z, and is usually estimated from the Wentzel equation:

$$w = Z^4/(Z^4 + c)$$

The fraction of excitations which gives Auger emission is $(1 - w)$. The constant c is about 10^6 for K-shell excitations, 10^8 for the L shell and 1.2×10^9 for the M shell. This means that the yield of X-rays is small from elements of low atomic number, while the converse is true for Auger electrons. To give an example, from carbon ($Z = 6$) we would expect 0.1 per cent of excited atoms to give a K X-ray, while 99.9 per cent would give an Auger electron of some type. On the other hand, from zirconium, 72 per cent of the emission would be K X-rays while the other 28 per cent is distributed among various Auger electrons.

The above arguments suggest that it would be better to use Auger electrons for analysing elements lighter than about zinc and X-rays for heavier elements. However life is not this simple, since X-ray detectors are currently smaller, cheaper and more easily fitted to a TEM than Auger spectrometers. Further consideration shows that it would, in principle, be better still to measure the energy lost by each electron passing through a specimen since this will enable us to detect each excitation, regardless of whether the excited atom subsequently emits an X-ray or an Auger electron. This argument indicates why there is such interest in EELS. However EELS will not come into its own, despite collecting virtually all the electrons which pass through the specimen, until the spectrometer can be made to perform as efficiently as an X-ray detector. Currently, most EEL spectrometers measure the energy of only a tiny fraction of the electrons entering them, whereas X-ray detectors are much more efficient, even at detecting very soft X-rays from light elements. Despite this handicap EELS is extremely useful for light-element analysis because, as shown above, there are many more characteristic energy-loss electrons than characteristic X-rays emitted from specimens containing light atoms.

The state of the art in 1987 is thus that only a very few highly-specialized transmission microscopes have an Auger spectrometer. Most TEM analysis of light elements is done by EELS or by X-ray analysis. Recent developments in both techniques include parallel detection for EELS and 'windowless' operation for X-ray detectors. These improvements, when they become widely available, will make the techniques more suitable for quantitative analysis of light elements and more accessible to the 'ordinary' user. In the remainder of this book we devote two chapters to each technique in order to describe how to use it in both a qualitative and a quantitative manner. In a final chapter we compare the relative strengths and weaknesses of the two techniques, which few microscopists are yet lucky enough to have available on the same instrument.

Windowless energy-dispersive X-ray analysis

The immediate benefits of windowless EDX (WEDX) for light-element analysis are evident in the spectra shown in Fig. 2.1. Without a beryllium window, the B_K and Ti_L peaks can be seen clearly. However, before we can interpret the details of the spectra we must consider the nature and performance of the detector itself.

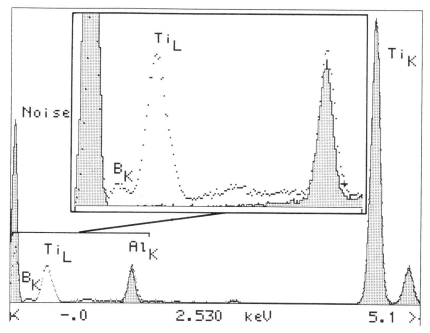

Fig. 2.1. A spectrum from titanium boride (TiB_2) with the window present (shaded spectrum) and absent (dotted spectrum).

A conventional Si(Li) X-ray detector is shown schematically in Fig. 2.2. The lithium-drifted silicon crystal must be cooled to liquid-nitrogen temperature in order to reduce its conductivity and to slow down the out-diffusion of the lithium dopant. The crystal must be biassed and in order to achieve this, a thin metallic contact layer (usually gold) is deposited on the front face. The cold crystal needs to be protected from vapour-phase contaminants which could condense on it, and

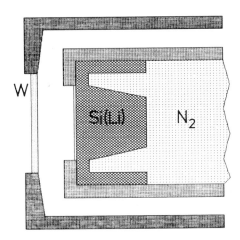

Fig. 2.2. A schematic diagram of a Si(Li) detector with a window (W). The crystal is cooled from the back by liquid nitrogen (N_2). The space between the crystal and the window is evacuated before the window is sealed into position.

from light which would trigger the detector. Both these aims are commonly achieved by evacuating the region around the crystal and sealing this cavity with a beryllium window.

Both the window and the gold contact layer unfortunately absorb incoming X-rays, and are therefore made as thin as possible. However the thinnest practicable Be window is several micrometres thick and absorbs soft (low-energy) X-rays so strongly that K lines from elements lighter than sodium are virtually undetectable, as are L or M lines with energies below 1 keV. Reference to Appendix 1 will show how many potentially useful X-ray lines are therefore inaccessible with a conventional detector.

The various approaches to extending the usefulness of EDX to lighter elements have all concentrated on allowing more soft X-rays to reach the active Si(Li) crystal and thus have a chance of being detected. The gold contact layer cannot be made any thinner (currently about 10-20 nm) without losing its effectiveness. The only remaining technique is to remove the beryllium window or to replace it with a film which absorbs fewer X-rays. This has led to the development of truly windowless or ultra-thin-window (UTW) detectors. Both these types are sometimes called 'windowless'.

Figure 2.3 shows the detection efficiency of an average Si(Li) detector with a conventional Be window, an ultra-thin window, and no window. It is clear that no detector is perfectly efficient for X-rays of energy less than 1 keV, although in absolutely ideal conditions a signal can be detected from beryllium K X-rays. However for boron K, the softest X-ray which is *easily* detectable, the overall detector efficiency can exceed 10 per cent with no window and is still a few per cent even with the ultra-thin window. As this figure shows, the detector fails

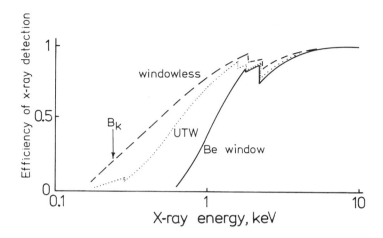

Fig. 2.3. The efficiency of three typical detectors (1 = all X-rays detected). The position of the boron K line (B_k) is indicated. ———, Conventional Be window; ·········, ultra-thin window; ------, no window.

to register all X-rays which fall on it. Part, but not all, of this effect is owing to absorption in the gold contact layer. The remaining loss is often attributed to the presence of a 'dead layer' at the surface of the silicon detector. However this is more accurately considered as a region of incomplete charge collection, from which all the electron-hole pairs created by the incoming X-ray photon do not contribute to the pulse of charge which leaves the detector. This results in the smearing out of the X-ray peak towards low energies, often known as 'low-energy tailing'.

Before considering the detector in detail it is useful to define several terms used to describe features of the X-ray spectrum.

Figure 2.4 is a sketch of the low-energy region of a hypothetical spectrum. It shows a noise peak close to the energy zero, a very low-energy peak which is not fully separated from the noise and another well-separated peak which exhibits low-energy tailing. The Gaussian ideal shape of this peak is shown dotted and its full width at half maximum (FWHM) is indicated. The 'resolution' of a detector is usually quoted as the FWHM of the Mn Kα peak at 5.9 keV. Peak widths at lower energies are generally smaller but are dominated by the width of the 'noise' peak as the expression

$$w^2 = w_n^2 + fE$$

shows. w is the FWHM of a peak at energy E and w_n is the FWHM of the electronic noise. f is a constant incorporating the Fano factor, and has a value of about 2.5 if all the units are in electron-volts. Thus a detector with a w_n value of 90 eV would have a quoted 'resolution' at Mn Kα of 151 eV but a FWHM at C_K (282 eV) of only 94 eV.

8 Light element analysis in the T.E.M.

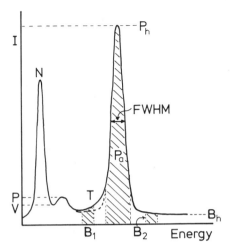

Fig. 2.4. A schematic low-energy spectrum showing the noise peak (N), a peak of area P_a showing low-energy tailing (T), and the peak-to-valley criterion (P/V). FWHM = full width at half maximum; P_h/B_h = peak height to the background height nearby; B_1, B_2 = backgrounds.

A second frequently-quoted parameter is the 'peak to background' ratio P/B. Some care needs to be taken in discussing this, since not all authors are referring to the same ratio. It is reasonable, for different purposes, to refer to the ratios of

(i) peak height to the background height nearby (this is P_h/B_h in Fig. 2.4);
(ii) peak area to an area of background of equal energy range, as shown by $P_a/(B_1 + B_2)$ in Fig. 2.4; or
(iii) peak height at Mn Kα to background height at 1 keV.

Many other combinations are possible, and there is even ambiguity about the definition of 'peak area' in (ii) and the reader should therefore examine closely the nature of P/B values quoted in the literature.

A further parameter often used to describe the low-energy quality of a spectrum is the peak-to-valley ratio P/V for a peak which is not fully resolved from the noise. This is also shown in Fig. 2.4.

2.1 The detector

There are many variables associated with the detector itself and its installation on the TEM. It is not wise to assume that all so-called 'windowless' detectors are the same. The major differences among detectors are likely to be associated with the area of active detector crystal, the position of the detector with respect to the specimen, and the nature of the 'window'. We will consider these points in turn, starting with the detector crystal. It is clear from simple geometry that, at a given position, the larger the detector the greater the fraction of emitted X-rays it will intersect. Since when analysing a thin specimen we are almost always in need of as much signal as we can get, a large detector would appear to be preferable. Commonly-available detectors have nominal areas of 10 or 30mm^2. However before choosing a 30 mm^2 detector, it is worthwhile to check its performance at low X-ray energies, since the higher capacitance associated with the larger detector area can introduce extra electronic noise (i.e. higher w_n) which can reduce the peak-to-background ratio and make the separation of similar energy peaks more difficult.

There are a variety of considerations which determine the best position for the detector on the microscope. The two most common configurations are often called 'low' and 'high' take-off-angle, and are shown schematically in Fig. 2.5. Because of the design of modern condenser/objective lenses it is usually possible to place a low take-off angle detector closer to the specimen than is possible with a high take-off angle. For a given area of detector crystal, the X-ray count rate for a high take-off angle detector will therefore be lower, maybe by a factor of 2 or 3. The continuum X-ray background (Bremsstrahlung) which lies beneath the peaks of the spectrum is more intense at low angles. The P/B (however defined) ought therefore to be better at high take-off angles. Early results with high take-off angle detectors suggest however that this theoretical advantage has not yet been realized in practice, perhaps because of interference by back-scattered electrons.

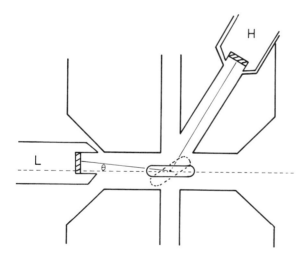

Fig. 2.5. Typical low-angle (L) and high-angle (H) positions for a detector. To use the low-angle detector the specimen usually needs to be tilted to the dashed position.

A crucial advantage of high take-off angle for many workers is the fact, illustrated in Fig. 2.5, that it is not generally necessary to tilt the specimen holder to 30° or so in order to collect an X-ray spectrum. This has three great advantages:

(i) the risk of the analysed area lying in the shadow of a grid bar or part of the specimen holder is much reduced, so a greater fraction of the visible area is available for analysis;
(ii) the absorption path which the emerging X-ray travels within the specimen is reduced, and thus quantitative analysis is more accurate (see Chapter 3); and more contentiously,
(iii) the range of tilt, and hence diffraction conditions available, is increased if the specimen does not have to be in one closely-specified orientation for analysis.

It appears too that on some installations the number of 'spurious' peaks such as Fe, Cu or Ti (which are seen in many spectra) is reduced for high take-off angle detectors.

Many detectors are mounted on a track which permits their distance from the specimen to be controlled. For most purposes the analyst needs as large a signal as possible, and therefore would want the detector as close to the specimen as possible. However it is possible, for example when analysing a 'thick' specimen, to flood the detector with X-rays or with back-scattered electrons and it is then convenient to be able to reduce the solid angle which the detector subtends by withdrawing it. Since it is also necessary to protect a windowless detector, at several installations the detector can be withdrawn behind a gate valve so that it can be isolated completely from the microscope column. This makes removal of the detector easier since the column vacuum need not necessarily be breached and permits, for those microscopes which need it, bake-out of the column.

Further protection for the detector in its normal working position may be needed to keep out back-scattered electrons. On some installations this is a particular problem when low microscope magnifications are used and the beam hits the effectively solid grid bars or thicker regions of the specimen. The effect of a large dose of back-scattered electrons can be to temporarily damage the detector so that it cannot be used effectively for some hours. It is therefore becoming increasingly common to install some type of safety device which shields the detector when low magnifications are selected.

Many detectors are fitted with a turret mechanism which enables the user to insert or remove a Be window. Either this, or a gate valve, or both, is clearly necessary so that air can be admitted to the column without the risk of ice forming on the detector. Some detectors offer the choice between a Be window and no window at all, while an increasing number use an ultra-thin window of Mylar or aluminium. Since the bare Si(Li) detector is sensitive to light, it is preferable to provide an opaque shield and organic film ultra-thin windows are therefore generally coated with a 10 nm film of aluminium for this purpose. It may not be obvious why light should be a problem behind the thick walls of the TEM. There are however two potentially rather bright sources: the specimen may cathodoluminesce and the thermionic electron gun gives off light which can be specularly reflected into the detector by a polished specimen. The latter effect has been detected in our laboratory by examining the 'spectrum' detected by a truly windowless detector when the filament was on but wildly misaligned so that light photons, but no electrons, were hitting the specimen. The effect of light on the X-ray detector usually appears as a degradation of the resolution, visible as a broadening of every peak.

To summarize this section;

- Windowless detectors may have interchangeable windows which may offer ultra-thin window or truly windowless operation.
- The detector(s) may be mounted at low or high take-off angles.
- The detector–specimen distance may be under the user's control.
- Detectors need protection against light and back-scattered electrons.

2.2. The electronics

The main components of the electronics chain associated with a Si(Li) detector are shown in Fig. 2.6. Although for most purposes microscopists can get away with considering these as 'black boxes' we do need to note a few of the factors which affect light-element analysis. The special feature of light-element analysis in the TEM is that we must detect and count very small signals (e.g. 185 eV for a boron K X-ray) in the presence of very large signals (e.g. 100 000 eV Bremsstrahlung X-rays). The electronics must therefore cope with a high-energy X-ray, which gives a large charge pulse, and recover very quickly to distinguish a very small pulse from the electronic noise inherent in the system. Excellent design of the FET and its head amplifier and the pulse-processing unit in modern commercial analysers have made this possible but it may still be necessary to use different operating conditions for the most efficient detection of soft X-rays from those in day-to-day use for conventional analysis.

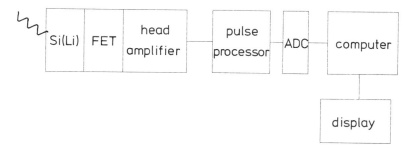

Fig. 2.6. The components of the electronic chain between the detector and the analytical computer. FET = field effect transistor, usually attached to the back of the detector crystal; ADC = analogue-to-digital convertor.

The three most accessible variables are the discriminator setting, the time constant of the pulse processor and the threshold of the analogue-to-digital convertor (ADC). The discriminator setting tells the pulse processor at what size of pulse it can believe that the pulse is 'real' and should not be discarded as noise. This does not guarantee that a pulse just above this size *is* real, but it does prevent the pulse processor having to waste time working on a lot of noise pulses. The time constant of the pulse processor dictates the efficiency with which the processor can control the size and shape of the pulse and thus enable the ADC to determine the X-ray energy. The longer is the processing time, the better will be the efficiency of detection of low-energy X-rays. However, the processor will be able to handle fewer pulses per second, so the count rate must be kept lower. For the current LINK pulse processors, for example, a time constant of 40 μs is suitable for light-element analysis, limiting the count rate to about 1500 cps. For conventional analysis, a time constant of 20 μs would normally be used, permitting the count rate to approach 3000 cps before the dead time becomes too large and the spectrum resolution is degraded.

In practice, for TEM-based analysis, it is rare to be able to work with a count rate in excess of 1500 cps so the use of a long time constant does not impose a real limitation.

The ADC threshold determines which pulses from the pulse processor unit (PPU) are actually measured and subsequently displayed. It is usually used to cut out all pulses below a specified energy and therefore its use is dangerous if we need to examine the very lightest elements. An example of how the threshold could be used to give a misleading impression of an excellent boron peak is shown in Fig. 2.7. With the appropriate threshold setting, a noise peak should be visible (see the dotted spectrum in Fig. 2.7). It may not be centred around the zero-energy position but at 20 to 30 eV. The presence of the noise peak is detected by some types of software and may be subsequently removed before the spectrum is processed further.

Each day, and after any change has been made to the processor parameters, it is advisable to check the calibration of the detector/processor combination. The way in which calibration is carried out differs slightly from system to system but the objective is always to ensure that the energy which the computer attributes to each channel of the spectrum is correct. It is usual for light-element analysis to use 10 eV per channel. The system must then be calibrated using a well-known spectrum, for instance from copper. The processor zero and gain settings must be adjusted so that a known channel (say number n) is receiving the zero energy signal and the channel $n + 8040/10$ is receiving the peak of the Cu Kα signal.

If the analyser generates and displays a reference peak (sometimes called the 'strobe peak') this can be used to set the zero-energy calibration but should be suppressed (by displaying or storing it elsewhere in the computer memory) before analysis is started.

2.3. Escape and sum peaks

There are two effects, familiar to electron microprobe analysts, which occasionally give rise to small spurious peaks in the spectrum. *Sum peaks* arise if the count rate

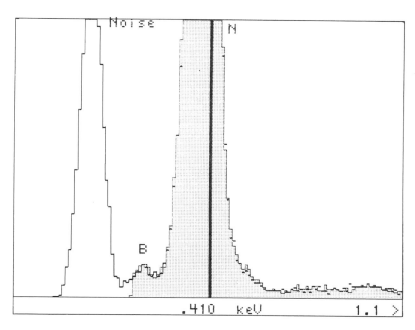

Fig. 2.7. A small boron peak accentuated by setting the threshold to give a sharp, low-energy cut-off (shaded spectrum). The dotted spectrum shows the effect of reducing the threshold to reveal the noise peak.

is large enough that there is a significant probability of two X-rays entering the detector at essentially the same time. The detector then recognizes these two pulses as a single pulse whose energy is the sum of their separate energies. A small peak may therefore appear at an energy greater than that of a major peak in the spectrum. Since count rates in X-ray analysis of thin specimens are generally rather low, sum peaks are rarely a problem. The only time a sum peak is likely to be encountered is during the analysis of a thick part of a specimen with one dominant element. If there is a sum peak it will then appear at an energy twice that of the main peak.

Escape peaks are more likely to be seen in light-element analysis. They arise from the fluorescence of a Si_K X-ray from the detector by the incoming X-ray which is to be detected. Such fluorescence occurs quite frequently and *if the Si_K X-ray then escapes from the detector* the energy detected will be lower than that of the arriving X-ray by 1.74 keV, which is the energy of a Si_K. This may then lead to the appearance of a small peak 1.74 keV below a major peak in the spectrum. Since the soft X-rays which are useful in light-element analysis are stopped close to the surface of the Si(Li) detector, the probability of escape of a Si_K X-ray is higher than it is during the detection of heavier elements. Analysts must therefore always be on the lookout for escape peaks, such as that shown in Fig. 2.8.

14 *Light element analysis in the T.E.M.*

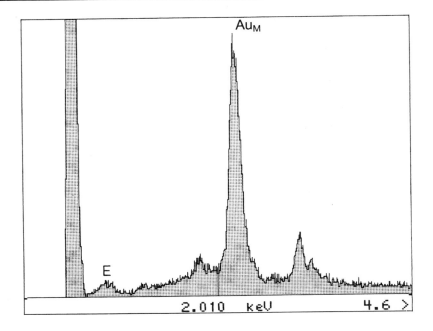

Fig. 2.8. An escape peak (E) below the main (M) peak from a specimen of gold.

2.4. Performing an analysis

Obviously much of the procedure for qualitative analysis of light elements is the same as for conventional analysis. In the following checklist the special precautions to be considered when looking for light elements are emboldened.

- Load specimen in analytical holder (e.g. beryllium)
- Select kilovoltage
- Select region of interest
- Select magnification
- Set up appropriate specimen orientation (tilt towards detector?)
- **Set processor time constant and threshold levels**
- **Select energy calibration; often 10eV/channel is best**
- **Check discriminator settings, and remove 'strobe' if applicable**
- **Check microscope vacuum level, if acceptable**
- **Open gate valve or rotate turret to UTW or windowless position**
- Analyse
- **Close window or withdraw detector behind gate valve**

Periodically it is wise to check that no ice or contamination layer has built up on the detector. This needs to be checked even if the detector has a UTW since windows can leak or be damaged. This should be done by inserting a familiar standard specimen, which is kept for this purpose. There are a number of possibilities: A gold specimen should give no oxygen signal, the presence of which is a good indicator of an ice layer. A nickel specimen, or other transition metal, has a convenient L line at low energy (see Appendix 1) and the ratio of the intensities of the L and Kα lines can be determined. Build-up of a contamination layer will be reflected in a decrease of this intensity ratio since the soft L X-ray will be absorbed strongly whereas the higher-energy K line will be relatively unaffected. Figure 2.9 shows the effect of icing on a spectrum of nickel oxide, NiO. Notice that not only is the Ni_L/Ni_K ratio much reduced by icing but the O_K/Ni_L ratio is also substantially altered.

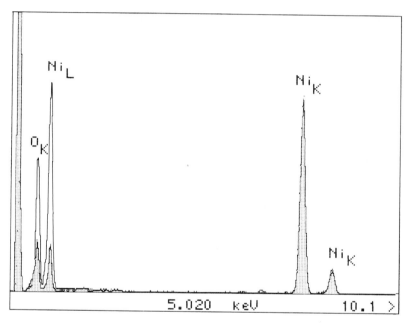

Fig. 2.9. Two spectra from the same specimen of nickel oxide, before (dotted spectrum) and after (shaded spectrum) icing of the detector. The Ni_K peaks are the same height but all the low-energy peaks are much reduced. Notice that the O_K peak is not reduced as much as Ni_L because an oxygen signal is being fluoresced from the ice.

If you should find that the detector is contaminated, then it should be cleaned. This is usually a job for the manufacturer, but if the layer is only of ice, it is tempting to reason that warm-up of the detector to room temperature will soon remove it. It is true that this should not harm the detector (as long as the bias voltage is switched off first) but consider where the water will go to! More than one 'clean' microscope column has been spoilt by water vapour from a detector being 'cleaned'.

16   *Light element analysis in the T.E.M.*

2.5. Qualitative interpretation of the spectrum

Full quantitative interpretation of the spectrum will be considered in more detail in Chapter 3 but because of the possible complexity of windowless spectra it is worth running over the steps involved in even a qualitative interpretation. The logical steps are:

1. Identify the position (i.e. the energy) of any obvious peaks.
2. If possible, identify them by reference to Appendix 1, bearing in mind the possibility of sum or escape peaks.
3. If this preliminary identification suggests an L line, then check for the presence of the equivalent K lines at higher energies.
4. If the line is apparently an M line, check for L lines at higher energies.
5. Consider apparent shoulders on the side of the 'easy' peaks similarly.
6. Consider what lines could be hidden by the peaks you have identified.
7. Consider whether absorption in the support film or in the specimen could have reduced the intensity of possible peaks. Perhaps you should have done this before collecting the spectrum!

2.6. Mapping

One of the most useful ways of displaying analytical information is as a map showing the spatial distribution of one or more elements. This is easy to arrange on any instrument with scanning facilities. The major requirement is that a signal proportional to the local concentration of the element can be used to modulate the brightness of the display screen. Most software for X-ray analysis permits the setting up of several 'windows' in any spectrum. Figure 2.10 shows carbon and nitrogen maps of some particles of boron nitride (BN) on a carbon film. Windows were set around the carbon and nitrogen peaks in the spectrum, and in a 'background' region. The computer output the count from each of these three groups of channels every time the scanning beam moved to a new pixel. In this way two 'elemental maps' were built up simultaneously. In producing the maps shown in Fig. 2.10 the counts in window 3 have been subtracted from the counts in regions 1 and 2, thus achieving a crude form of background subtraction. It can be seen that the carbon map is fairly uniform, with a slight indication of a locally higher concentration around the particles, while the nitrogen is clearly only to be found within the BN particles. The other two images in Fig. 2.10 show the scanning transmission image and a 'smoothed' version of the nitrogen map.

There are a number of pitfalls for the unwary analyst when light elements are mapped. The local intensity of a map cannot be taken to indicate the local concentration of the element being mapped, since the signal (which is what is being displayed) is approximately proportional to the local thickness. In addition the effect of absorption may be to mask the presence of light elements on the underside of the specimen. A further problem which is more acute with light elements is

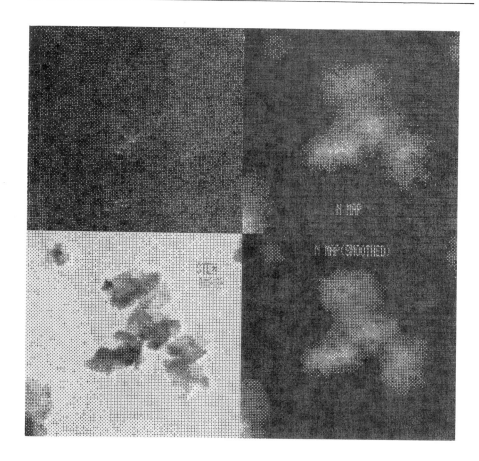

Fig. 2.10. X-ray mapping of a cluster of boron nitride particles. The images are the STEM image (*lower left*), carbon map (*upper left*), nitrogen map (*upper right*) and smoothed nitrogen map (*lower right*).

the difficulty of setting windows on peaks which are close together and tend to overlap. Also since the background varies rapidly at the low-energy end of the spectrum it is difficult to set an appropriate background window.

Finally, we must emphasize the low accuracy and detection limit which are set by the poor statistics of an X-ray map from a thin specimen. The maps shown here consist of 128 × 128 pixels (i.e. 16 384 analysed points). They were collected at 200 ms per point and therefore took about 55 min. to create, even after the microscope and specimen had been set up. Despite this long time, the most intense pixels contain less than a hundred counts. For all the thinner regions the statistics are worse. Nonetheless, maps of two or more elements are one of the best ways of showing the relationships between elements in a complex sample.

WEDX: Quantification and limits

In Chapter 1 we showed that the ionization cross-section for light elements is higher than for heavier elements, although the fluorescence yield is lower. Thus an electron beam carrying a particular current will excite light atoms more efficiently than heavy atoms but fewer of them will relax by emitting an X-ray. On balance these effects almost cancel out and this means that the *production* of X-rays from light elements is not a particular problem, as Fig. 3.1. shows.

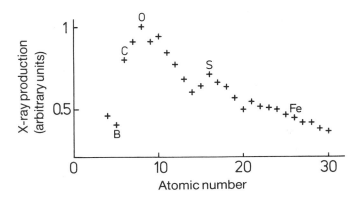

Fig. 3.1. X-ray production (i.e. excitation efficiency multiplied by fluorescence yield) for the K line of the elements Be to Cu. The vertical scale is arbitrary, normalized to unity for oxygen.

The two main factors which make light-element X-ray analysis difficult are the inefficiency of the detector and the tendency for the X-rays of interest to be absorbed in the specimen. We have considered the first of these points in Chapter 2 but in order to quantify light-element analysis we must consider absorption in some detail, and this forms a major topic of this chapter. First however, let us consider the approaches to quantification which are common in the physical and materials sciences.

All approaches to quantification start with three essential steps, designed to deduce the number of characteristic X-ray counts for each element of interest in the collected spectrum. These numbers are essentially integrated peak areas, as Fig. 3.2. shows, but are generally referred to as intensities, I. The steps are:

(1) identification of the X-ray peaks of interest;
(2) removal of 'background' from beneath each peak;
(3) deconvolution of overlapping peaks.

WEDX: Quantification and limits 19

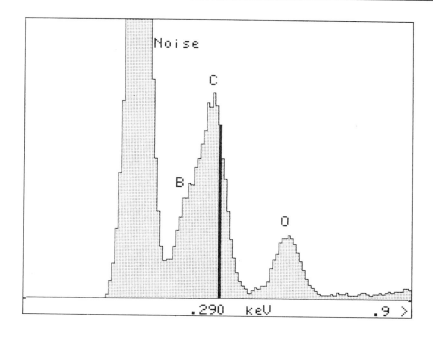

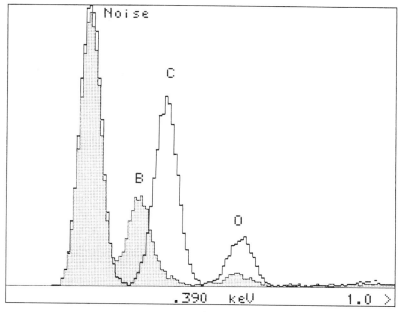

Fig. 3.2. The composite peak from boron and carbon (a) is shown deconvoluted into the original almost-Gaussian peaks whose areas are needed for quantitative analysis (b).

These three steps are frequently accomplished by commercial EDS software packages and the details of the processes may be hidden from the analyst. This is regrettable since it can mean that the limitations of the processing are overlooked and spurious accuracy may be attributed to the resulting peak areas. Other handbooks (e.g. Morgan 1985) deal with these techniques and since the approach is not different in principle for light-element peaks, we will not deal with it in great detail, although we urge the reader to become familiar with the technique used in his or her laboratory. The important differences which will be noticed when performing light-element analyses are that peak overlaps are more likely to be encountered and background subtraction is not as straightforward since the background varies more strongly in the low-energy region of the spectrum. In addition it may prove more difficult to obtain 'standard' peak profiles for light elements, whether they be experimental or calculated, since elemental standards are frequently impossible to use, for example for gaseous elements.

After the initial three steps the approaches of the biologist and the materials scientist generally diverge. The intensities of each peak (I_A from element A, I_B from element B etc.) are interpreted in significantly different ways:

'Materials' approach	'Biological' approach
Ratio method	Continuum method
Concentration, C_A expressed as weight per cent	Concentration, C_A, expressed as mass fraction or in molar terms as mM/kg or mM/litre
Intensity ratio from two elements used to deduce the concentration ratio via a 'K-factor' which must be known for the instrumental conditions	Concentration deduced from ratio of peak intensity to intensity of the continuum, W, (i.e. background), over a specified energy range
$\dfrac{C_A}{C_B} = K_{AB} \cdot \dfrac{I_A}{I_B}$ (3.1)	$C_A = k \cdot \dfrac{I_A}{W}$ (3.2)

These expressions are very simple and work quite well if the specimen is thin enough for the absorption of emitted X-rays to be neglected. This is the so-called 'thin-film criterion'. However for very soft X-rays, absorption becomes significant when the specimen is still extremely thin, for example for the determination of boron in aluminium the thin-film criterion breaks down at specimen thicknesses greater than 2 nm. Thus the factors K_{AB} and k, which can be assumed to be constant for many conventional analytical purposes, must be modified to take account of absorption if light elements are to be analysed.

3.1. Determination of K factors

The accurate determination of either K_{AB} or k requires the use of standards. The detailed approaches are treated in other handbooks but we still emphasize here the aspects which are of particular importance when dealing with light elements.

For biological analysis, a standard (std) of known mass–thickness, containing the element of interest in concentration C_{std}, is needed. It need not be exactly the same thickness as the unknown sample. By applying (3.2) twice (see above Table), once for the standard and once for the unknown, and assuming that the value of k remains the same, the concentration in the unknown, C_A, can be determined from:

$$C_A = C_{std} \ (I_A/W)_{unknown}/(I_A/W)_{std} \qquad (3.3)$$

For light-element analysis, the intensities I_A should be corrected for absorption, as detailed below, before applying eqn (3.3). Strictly, it should also be necessary to correct the continuum measurements, W, for absorption although this is less frequently attempted since W is generally measured at a high X-ray energy where absorption is unimportant.

In order to apply eqn (3.1) to 'materials' specimens, standards are also required. In this case it is necessary to find samples of known composition which contain at least two of the elements of interest. The K factor K_{AB} can be determined from eqn (3.1) if measurements of I_A and I_B are made from a standard in which C_A and C_B are known. Again it must be emphasized that for light-element analysis, the measured I values must be corrected for absorption before K_{AB} is calculated, otherwise the K factor will be applicable only to specimens of one thickness. Most of the remainder of this chapter is devoted to the absorption correction and the determination of sample thickness, which is a necessary precursor to it.

The choice of standards is particularly problematical for light elements. In principle the K factor K_{XY} which relates any two elements X and Y can be deduced from a series of other K factors which eventually relate both X and Y to a common element. The simplest example of this would be:

$$K_{XY} = K_{XA}/K_{YA} \qquad (3.4)$$

There are however a rather limited number of well-characterized, beam-stable, homogeneous, light-element-containing, easily-thinned materials. Some comments on the possibilites are as follows: For boron, there are a number of transition-metal borides such as TiB_2, but preparing them is not easy and absorption is so serious that the region used for analysis will need to be overhanging the edge of a hole in any support film. The elimination of an overlapping carbon peak will still generally be necessary and may limit the potential accuracy. Stable carbides also tend to be formed with elements in the middle of the Periodic Table and similar problems will be encountered with support films. In addition, contamination frequently provides a time-dependent carbon signal.

Nitrogen is quite difficult to standardize. The most readily-available stable compound is often boron nitride but this is not useful as a standard because it does not relate nitrogen to a heavier element. Several potentially-suitable nitrides exist but none has found universal favour as a standard. Silicon nitride is probably the best candidate.

Suitable oxides are more plentiful and include Al_2O_3, SiO_2 and TiO_2. The problem with oxygen is more likely to be associated with the uncertainty of the necessary assumption that the unknown specimen is homogeneous throughout the analysed volume than with the K factor.

3.2. Absorption of X-rays in the specimen

We will now consider the geometry of absorption in a thin specimen. In the general case of a specimen which may be tilted towards the detector by an angle ϕ, where the detector is at an elevation θ, the path length which an emerging X-ray has to travel is shown by x in Fig. 3.3. If z is the depth in the specimen *along the beam direction* at which the X-ray is excited, then

$$x = z \cos(\phi) \operatorname{cosec}(\theta + \phi) \tag{3.5}$$

Absorption is described in terms of a reduction in X-ray intensity via the usual expression:

$$I = I_0 \exp(-\mu\rho x) \tag{3.6}$$

where ρ is the specimen density and μ is the *mass absorption coefficient* for the appropriate X-ray, which is defined as the linear absorption coefficient divided by the density and is usually quoted in cm^2/g; μ needs to be the average value for the composition along the path length and assuming that the specimen is homogeneous over this distance (which it might not be), this is given in terms of the elemental mass-absorption coefficients by:

$$\mu = C_A\mu_A + C_B\mu_B + C_C\mu_C \ldots \tag{3.7}$$

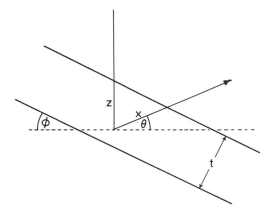

Fig. 3.3. The geometry of absorption. x is the path length which an X-ray must travel to a detector at elevation θ after being excited at depth z. The specimen is tilted ϕ from the horizontal.

It is clear that strictly we need to know the composition of the specimen before calculating μ, but in practice it is often accurate enough to calculate a first estimate of each C from the experimental (uncorrected) I values, using conventional k factors. Only if the absorption correction turns out to be more than about 10 per cent is it necessary to iterate to a better solution by recalculating the mass absorption coefficients and performing the correction again.

Since it does involve a little algebra it is worth working through the calculation of C from I in a system containing more than two elements. Taking a three-element sample as an example, we must assume that

$$C_A + C_B + C_C = 1$$

then

$$\frac{C_C}{C_B} = K_{CB} \cdot \frac{I_C}{I_B}$$

$$\frac{C_A}{C_B} = K_{AB} \cdot \frac{I_A}{I_B} ;$$

eliminating C_A and C_B from these three equations we find that

$$C_A = I_B/(I_C K_{CB} + I_A K_{AB} + I_B)$$

In order to calculate the total reduction in intensity owing to absorption in a thin specimen we need to sum the effects at all depths. If we assume that X-ray generation is equally likely at all depths we can do this by integrating eqn (3.6) for values of z from 0 to $t/\cos(\phi)$. There are two assumptions implicit in this approach. One is that the physics of electron scattering does give equal probabilities of excitation at any depth, and this is usually considered to be reasonable. The other assumption is that the specimen is homogeneous throughout its thickness. This is more questionable, but its validity can only be assessed by the scientist whose specimen it is. If we accept the assumptions, the integration gives:

$$I = I_0 \frac{[1 - \exp(-\mu\rho t \cosec(\theta + \phi))]}{\mu\rho \cos(\phi) \cosec(\theta + \phi)} \quad (3.8)$$

where I_0 is the intensity generated per centimetre of specimen thickness. It is immediately evident that in order to calculate the effect of absorption we must know the thickness of the specimen, t. This is a major problem for many specimens, but before considering the techniques available to measure it, we should define very carefully what we mean by specimen thickness. t is shown in Fig. 3.3 as the perpendicular separation of the (notionally parallel) surfaces of the specimen. However if, as is quite likely, the measured specimen thickness refers to the thickness along the beam direction with the specimen in the orientation used for analysis, then the $\cos(\phi)$ term should be omitted from the numerator of eqn (3.8).

It is obvious that the magnitude of the absorption effect depends on μ, ρ, t and the specimen/detector geometry: All these parameters appear within the exponential in eqn (3.8) and the absorption effect may therefore be very sensitive to variations in their values. During the analysis of any point we can assume that all the parameters are constant except for μ, which depends on the energy of the characteristic X-ray being considered. In order to determine the correction factors K_{AB} or k described above it is necessary to know the mass absorption coefficients for all the X-rays of interest in all the elements of the specimen. This poses a serious problem: Several workers have either measured or calculated mass absorption coefficients and a number of tabulations are available. There is generally good agreement between these values for the absorption of X-rays of energy >1 keV in all elements, and fair agreement even for very soft (e.g. light element) X-rays in light elements. However there is a huge range of values published for the absorption of soft X-rays in heavy elements. This is a pity, since many analytical problems involving light elements inevitably require their determination in the presence of elements of medium-to-high atomic number.

Appendix 2 lists some published values of mass-absorption coefficients for the seven light-element K lines from boron to sodium in a range of light and heavy elements. The figures in bold type are those of Henke and Ebisu which are probably the most widely used. However in order that the reader can get some idea of the possible error in these values, a limited number of values from two other sources are also quoted. It is evident that the agreement is encouraging for absorption in light elements but the disparities are alarming for absorption in elements with atomic number above about 20.

It is worth using some of this data to assess the magnitude of the absorption effect in some specific analytical situations. In Table 3.1 is shown the fraction of the excited X-rays which will be absorbed by a homogeneous specimen before reaching a typical low or higher take-off angle detector. These figures do not take account of absorption in the detector, which would have to be superimposed.

Table 3.1. *The fraction of K X-rays absorbed in specimens of three thicknesses with detectors at a low take-off angle ($\theta = 7°$) with the specimen tilted ($\phi = 30°$ towards detector) and at high ($\theta = 60°$) with the specimen untilted ($\phi = 0$). The mass absorption coefficients were taken from Henke and Ebisu (see Appendix 2)*

System	Density	Thickness					
		10nm		50nm		200nm	
		low	high	low	high	low	high
B in Fe	7.86	0.15	0.11	0.51	0.41	0.85	0.79
C in polymer	1.00	0.00	0.00	0.01	0.01	0.06	0.04
C in $Cr_{23}C_6$	6.7	0.05	0.04	0.23	0.17	0.60	0.49
O in Al_2O_3	3.97	0.01	0.01	0.06	0.05	0.23	0.17
O in CdO	8.15	0.11	0.08	0.42	0.32	0.79	0.71
Na in NaCl	2.17	0.00	0.00	0.02	0.01	0.06	0.04

It is immediately clear that the absorption effect can rarely be ignored if light elements are to be analysed. Even for specimen thicknesses as small as 10 nm there may be a 5 or 10 per cent loss due to absorption, for example in the cases of C in

$Cr_{23}C_6$ or B in Fe. For thick specimens the effect may be huge (e.g. O in CdO at 200 nm) and a correction factor of about 5 [i.e. $1/(1-0.79)$] would need to be applied to the measured intensities in order to deduce the oxygen concentration. However, the exact value of such a large correction factor would depend crucially on the mass absorption coefficient, which in this case would be dominated by μ for O in Cd. As emphasized above, this is the least accurately known type of mass absorption coefficient. Currently the only way to determine a light element accurately is to avoid the necessity for using poorly-known mass absorption coefficients, that is by analysing only a very thin specimen. In the case of O in CdO, which has been used as an example, this probably means restricting the analysis to areas of thickness no greater than 10 nm. This is a serious restriction, not welcomed by most microscopists.

No improvement to the absorption problem can be effected by operating at higher electron energies, since the figures in Table 3.1 arise solely from the absorption of the emerging X-rays. Indeed the situation could in practice be worse at higher operating voltages since the microscopist can see through thicker regions and can easily be persuaded into thinking that a choice area for analysis is thinner than it really is. Although there may be many reasons for favouring higher-voltage microscopes, including a potential increase in the signal-to-background ratio, the X-ray analysis of light elements is not one of them.

The effect of detector take-off angle is clearly shown in Table 3.1. It was suggested in Chapter 2 that one of the advantages of a high take-off angle detector is the reduction in the absorption effect. It can be seen from the table that although this is true the magnitude of the improvement is not great. None of the comments in the preceding two paragraphs would need to be modified significantly to apply to a high take-off angle detector. On its own, the absorption effect does not generally provide sufficient reason for choosing a high take-off angle detector geometry.

The absorption effect is obviously very important if a quantitative analysis involving light elements is to be attempted. However it is also necessary to take absorption into account in some qualitative analyses. In particular, if the region to be analysed is at one surface of the foil or support film, then it becomes crucial to know whether it is at the surface nearer the detector or whether the emitted X-rays have to penetrate the whole thickness of the specimen or film before reaching the detector. This consideration becomes important when analysing particles or precipitates in foils or extraction replicas. The magnitude of the effect is much greater than that implied by Table 3.1 if the particle is on the 'wrong' side of the foil since *all* emitted X-rays are then subject to absorption, as Fig. 3.4 shows. Table 3.1 was calculated for the integrated effect of absorption of X-rays generated throughout the foil, some of which would have been near the surface and therefore absorbed only slightly.

As an example, consider the detection of C from $Cr_{23}C_6$ in a steel: If a carbide particle of diameter 50 nm was sitting on the detector side of a 100 nm steel foil, the loss of carbon signal to absorption would be that given in Table 3.1, say 0.23 for a low take-off angle detector. If the same particle was on the other side of the

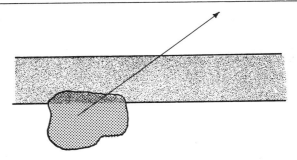

Fig. 3.4. X-rays from a particle beneath a support film must traverse an absorption path through the film.

foil, the C signal would be attenuated by an extra 100 nm of steel, and the total fraction absorbed would rise to 0.83. It is then quite likely that a carbon peak would not even be detected. It is therefore important to establish that the region chosen for analysis lies on the detector side of the sample, and for this purpose the easiest method, if it is available, is to examine the specimen in the SEM mode of a TEM/STEM instrument. If you cannot easily tell on which side the region lies, bear in mind that absorption may provide the reason for a deceptively-low signal.

3.3. Performing a quantitative analysis

In attempting a quantitative analysis involving light elements, the procedure for qualititative analysis summarized in Chapter 2 (p. 14) should of course be carried out first. Then the following additional steps are needed:

1. Collect spectra from several apparently-appropriate regions, if possible. Select those spectra which appear to have suffered least from absorption for further treatment. In most cases this means rejecting spectra with light-element peaks significantly lower than those in other spectra.
2. Estimate the local thickness at the region analysed, by one of the methods described below.
3. Determine the peak areas for each X-ray line of interest, using your locally-available software to remove the background and separate peak overlaps. These are then the I_A, I_B, I_C ... values.
4. Estimate the mass absorption coefficients for each line of interest in the specimen, using eqn (3.7).
5. Use eqn (3.8) to correct each value of I in order to compensate for absorption. The equation should be used in the form:

$$I_{OA} = I_A \cdot \frac{\mu^A \rho \cos(\phi) \operatorname{cosec}(\theta + \phi)}{[1 - \exp(-\mu^A \rho t \operatorname{cosec}(\theta + \phi))]} \tag{3.9}$$

for element A, where μ^A in this case is the mass absorption coefficient for A X-rays in the specimen.

6. Deduce the elemental concentrations from the absorption-corrected intensity values, using either the ratio or the continuum method, as appropriate. The k factors which are used for this correction (K_{AB} or k in the discussion on p. 20) must be values appropriate to the thin-film case where absorption is negligible. In other words, when determining k factors for light elements on your instrument, always correct for absorption before citing the value.

7. If the absorption correction changed the intensity values by more than say 10 per cent, then return to step 4 and re-correct using mass absorption coefficients calculated from the new concentrations C_A etc. using eqn (3.7).

3.4. Methods of determining local specimen thickness

It should by now be obvious that the determination of the thickness of the region being analysed is very important if light-element analyses are to be made at all quantitative. There are many techniques which have been used to estimate thickness, each of which has its advantages and disadvantages. Not all methods are available in all laboratories, nor are all applicable to every type of specimen. We will therefore give a brief description of several methods which can be applied to a selected small region of a specimen, together with references to more detailed treatments of their application.

The ideal method would be applicable to any specimen, accurate, quick and easy to execute, and would give the thickness at exactly the point from which the spectrum was collected. No method known to us fulfils all these criteria! Some of the more widely-used methods are:

3.4.1. X-ray intensity

It is relatively easy to compare the intensities of a peak or selected area of background, chosen to be of sufficient energy to be unaffected by absorption, in two regions of the specimen or even in two specimens. If these intensities are measured under identical microscope conditions they give the *relative* thicknesses of the two regions. This can be very useful in deducing the thicknesses of several regions after one of them has been measured by a more absolute technique. It can also be used to measure the thickness if one of the regions, or specimens, is a standard. This is a technique sometimes employed by biological microscopists, who use a polymer film of well-characterized mass–thickness as a standard (e.g. Hall and Gupta 1979).

3.4.2. EELS plasmon ratios

If the analytical microscope has an EEL spectrometer then this quick and fairly accurate method can be used. As discussed later in Chapter 5, the low-loss region of an EELS spectrum contains substantial peaks resulting from plasmon scattering (see Fig. 4.8). Since the mean free path for such scattering is small, the intensity of these peaks is strongly dependent on specimen thickness. As shown in Chapter 5, the ratio of the intensity of the zero-loss peak (I_0) and the single-plasmon loss peak (I_1) is related to thickness, t, by the equation:

$$I_1/I_0 = t/\lambda \qquad (3.10)$$

where λ is the 'plasmon mean free path'. Strictly, λ depends on the elemental composition of the specimen, and on the geometry (e.g. collection angle) of the EELS detector. However if λ is determined by measuring I_1/I_0 for a few specimens of known thickness, the same value can be applied to other materials without an error greater than 10-20 per cent, as long as the spectrometer conditions are the same.

3.4.3. Convergent beam electron diffraction (CBED)

This well-established method is generally considered to be the most accurate way of measuring the thickness of a crystalline specimen, and accuracies of 3-5 per cent are generally quoted. It relies on the measurement of the spacings of fringes in two-beam convergent beam diffraction patterns, and is well described by Kelly *et al.* (1975) and Allen and Hall (1982). The main disadvantages of the technique are that locally the specimen must be a single crystal without too many defects and that it must be in a goniometer holder capable of being used to set up accurate two-beam conditions quite easily. In addition, not all microscopes have the appropriate lens geometry and cleanliness to permit convergent beam patterns to be displayed and photographed. However, where the microscope and operator skills exist this is an excellent technique.

3.4.4. Contamination spot method

This method, despite its many drawbacks, remains one of the most widely used because of its simplicity. The principle is straightforward: The electron beam is focused down to a small spot, as is probably necessary for analysis anyway, and is allowed to remain on the analysed region for long enough to leave a significant spot of contamination on both surfaces of the specimen. After the analysis this region is tilted through a known angle, the separation (S) of the spots is measured (Fig. 3.5), and the thickness is deduced by simple geometry. Despite the protestations of those with clean columns and cleaner specimens, most microscope/specimen combinations do give rise to contamination spots during the analysis of a 'point' and this technique therefore has the advantages of speed, simplicity and virtual universality. However it suffers several serious limitations of which the two most important are that it is not easy to define the edge of the spots in order to measure their separation and it tends to overestimate thickness because of the presence of pre-existing thin contamination layers or oxides. Few workers would expect an accuracy better than +30/−10 per cent using this technique. It is also likely to become harder to apply the technique as microscopes become cleaner and microscopists learn to keep their specimens free from contamination.

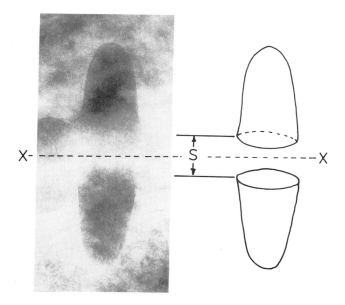

Fig. 3.5. A contamination spot and a diagram showing its idealized shape, from which the spacing S must be measured in order to determine the specimen thickness. Since the contamination spot was formed, the specimen has been tilted 45° about the axis X–X.

3.4.5. K/L intensity ratio

If the specimen contains a substantial fraction of an element of medium or high atomic mass it may be possible to compare the intensities of a K line and an L line from the same element, or an L line and an M line. Since the lower energy line will suffer absorption to a much greater extent than the higher energy line, this ratio contains information about the specimen thickness. Several workers have used this approach to determine local thickness and it has been shown, on steel specimens containing carbides, to give thickness values intermediate between those deduced by CBED and contamination spots (Rong *et al.*, 1984). The method does need calibrating for each microscope, since some of the K/L differences arise from the detector, but it offers another approach which may be useful for certain specimens.

3.4.6. Comparison of thickness measurement techniques

In the following table some comments on the relative merits of the techniques discussed above are offered. The techniques are listed in order of their probable accuracy:

30 *Light element analysis in the T.E.M.*

Technique	Comment
CBED	If the TEM is suitable. Only for fairly perfect crystals in a goniometer holder. Gives thickness of crystal, excluding surface layers of different material (e.g. oxide or contamination)
X-ray	Good relative technique. Absolute if calibrated. Applicable to any specimen
EELS plasmon ratio	If EELS available. Easy, once calibrated. Almost any specimen. Gives total thickness
K/L ratio	If appropriate elements present. Needs calibration
Contamination spot	If contamination is significant. Measures total thickness and still over-estimates.

3.5. Accuracy and detection limits

It is extremely difficult to make general statements about the potential accuracy and detection limits for analysis involving light elements. The sensitivity which is achievable will depend on the nature of the specimen and the microscope/detector combination and also on the geometry of the specimen. All that can be given is some advice on the considerations which should be taken into account when estimating detection limits or possible errors.

Conventionally, minimum detection limits are estimated by considering the statistics of detecting a peak superimposed on a randomly noisy background. A traditional analysis, which assumes the simplified situation shown in Fig. 3.6, would be developed in the following way: A peak P is assumed to be detectable if counting statistics of the background are truly random then $\sigma_B = B^{1/2}$ so the minimum detectable peak is of size $3B^{1/2}$. This is quite easy to estimate for a discrete

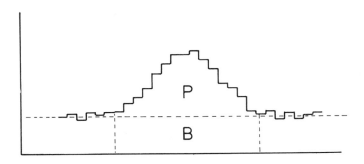

Fig. 3.6. The peak (P) and background (B) areas involved in estimating the minimum detectable concentration.

peak on an essentially flat background, and from a knowledge of the appropriate k factors can generally be converted into an estimate of the minimum detectable concentration.

However, the factors limiting the statistics of a light-element analysis often include deconvolution of overlapping peaks, the presence of a very non-flat background and the need for a large k factor. Estimating the errors in such procedures is not straightforward, so a single criterion for the visibility of a peak cannot be given. All we can state with confidence is that detection limits will always be worse than those estimated using the simple procedure outlined above.

Both accuracy and detection limits are affected by the chosen compromise between count rate, specimen thickness, counting time and contamination rate. In most real cases of light-element analysis these are not freely-selectable variables: Specimen thickness is limited by the need to avoid major absorption corrections; the count rate available is then likely to be very low; the counting time is also probably limited by the build-up of contamination, so a spectrum containing a large number of counts can rarely be collected. The detection limit must depend on how bad each of these factors is for the particular specimen/microscope combination on the day.

Particular elements also pose specific problems. Carbon is notoriously difficult to quantify, not because the signal is weak but because almost all contamination which builds up contains carbon and therefore the signal is likely to increase (spuriously) with time of analysis. Oxygen is also problematical because it is relatively easy to detect but tends to arise on every specimen because of the formation of oxides. In these cases, quantification is difficult because the assumption that the specimen is homogeneous throughout the analysed volume probably breaks down. The extraction of a boron peak gives rise to another set of problems. At this low energy (185 eV), the small peak must almost always be deconvoluted from a noise peak on the low-energy side and a carbon peak on the high-energy side, as illustrated in Fig. 3.7.

This chapter will close with an example of light-element analysis in which quantification has been carried as far as is practicable and an understanding of the subtleties of absorption is needed to deduce the solution. The spectrum shown in Fig. 3.8 is one of a series collected from an amorphous specimen of intended composition CuZr. The presence of oxygen is clear and if it is assumed that the oxygen is distributed homogeneously an absorption-corrected quantitative analysis can be performed. The appropriate K factors, K_{XFe}, for the instrument/detector combination were 1.176 for the Cu$_K$ lines, 1.35 for Zr$_L$ and 2.90 for O$_K$. This approach gave the results shown in Table 3.2.

The naïve interpretation of this data is that, since the composition does not change with thickness, the oxygen must be uniformly distributed throughout the specimen. However this is not correct: Because O$_K$ is strongly absorbed, the oxygen signal should at first increase with thickness and then reach a constant level when the signal from the bottom of the specimen is totally absorbed. This would occur at a thickness of about 500 nm. However if the uncorrected O/Cu ratio is plotted

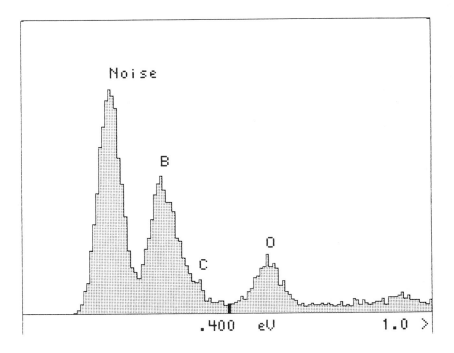

Fig. 3.7. A substantial boron K peak which is overlapped by the noise peak at lower energy and a carbon peak at higher energy.

against thickness it can be seen to fall very rapidly (Fig. 3.9). This behaviour would be expected if the oxygen was present in two layers of constant thickness, one at each surface. As the thickness is increased the contribution of the lower layer to the oxygen signal should decrease, again reaching a plateau at about 500 nm. Further experiments involving Rutherford back-scattering of He ions were able to

Table 3.2

Thickness (nm)	Weight per cent		
	O	Cu	Zr
46	29	31	40
73	30	32	38
85	37	28	35
131	34	29	37
148	30	32	38
150	32	31	37
193	30	31	39
218	30	31	39

WEDX: Quantification and limits   33

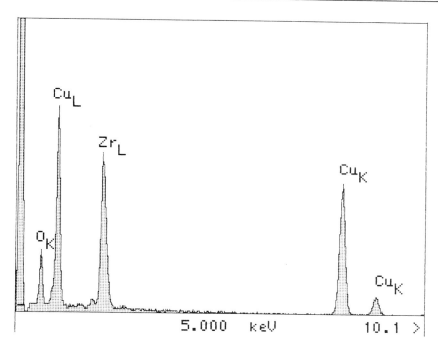

Fig. 3.8. A windowless spectrum from a specimen of nominal composition CuZr, showing a substantial oxygen peak.

confirm that this was indeed the case and the oxygen was confined to surface layers about 22 nm thick. Fig. 3.9 shows that the data fit much more accurately to the oxide-layer model than to a homogeneous distribution. This example illustrates the constant need to bear in mind the physical reality of the specimen rather than being lulled by the output of 'computer correction' programs.

34 *Light element analysis in the T.E.M.*

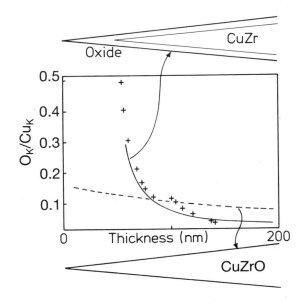

Fig. 3.9. Measurements of the ratio of O_K to Cu_K peaks as a function of specimen thickness. The data (+) compare well with a model of two oxide surface layers (———) but poorly with a model based on a homogeneous oxygen distribution (-------).

Electron energy loss spectrometry

An electron energy loss spectrometer (EELS) coupled to a TEM or STEM becomes a powerful microanalytical tool. In this chapter, the operation of the spectrometer, its coupling to the microscope and the optimization of the experimental parameters involved are explained first. Finally, the type of spectrum obtained and the information that it gives are examined qualitatively. An example of a typical spectrum is shown in Fig. 4.1 and it illustrates the three different regions observed: the zero-loss peak, the low-loss region and the higher-energy losses where the characteristic ionization edges are found.

4.1. Description of the spectrometer

There are now two types of EEL spectrometer commercially available, although at present the serial type is the one more frequently found in operation. However, the recent development of the parallel spectrometer with its many advantages is likely to lead in the near future to its gaining popularity. A description is given below of the main features of each type of spectrometer.

Figure 4.2(A) is a schematic diagram of the main components of the type of serial EELS which is currently commercially available. A magnetic spectrometer is used to separate the electron beam into its various energy components and the resulting distribution of electrons is then scanned, by varying the magnetic field, across a narrow slit so that at each moment electrons of a particular energy pass through the slit and the whole spectrum is collected serially. It is important to note that stray fields near to the spectrometer may affect this scanning of the electron spectrum causing the spectrum to apparently drift along the energy axis. In particular, it is necessary to avoid using chairs containing steel close to the spectrometer. The slit acts as an energy selector allowing each specific energy in turn to impinge on the scintillator/photomultiplier system. The resolution of the spectrometer is determined by the slit width since a certain range of energies $E - \Delta E$ to E will be allowed through, (see Fig. 4.2A). A factor limiting energy resolution could be spectrometer aberrations, if the spectrometer cannot focus a point on the object to a point in the plane of the slit. However, most spectrometers allow complete correction for second-order aberrations, which are the ones which would give the problems, enabling resolutions of fractions of an electron-volt to be obtained.

One of the diffficulties in energy loss spectroscopy is dealing with the large dynamic range in signal which has to be detected. For example, in the zero-loss peak, there may be several million counts per channel whereas at higher energies the number of counts in a channel may only be in the order of hundreds. In practice,

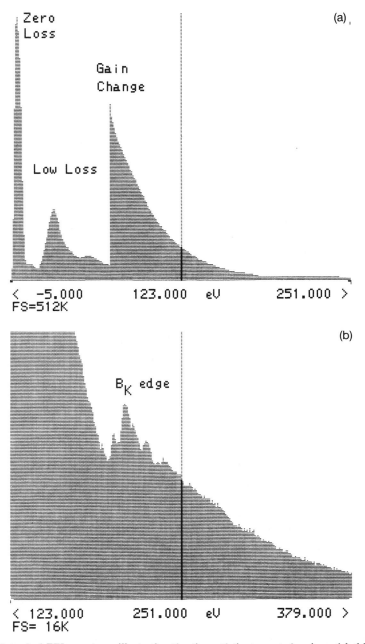

Fig. 4.1. A typical EEL spectrum illustrating the three distinct spectral regions: (a) this shows the zero loss peak and the low-loss region; (b) the higher energy losses showing the B_K edge. Note the change in scale.

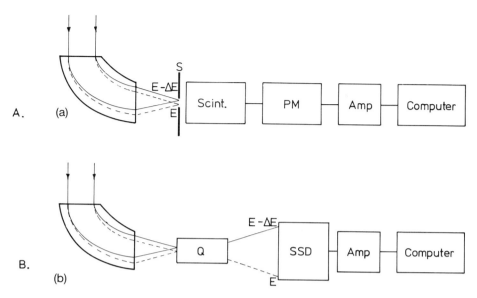

Fig. 4.2. (A) A schematic diagram of the main components of a serial EEL spectrometer: the slit S, scintillator (Scint), photomultiplier tube (PM), amplifier (Amp) and computer. (B) A parallel EEL spectrometer showing the main components: the quadrupole lenses (Q), solid-state detector (SSD), amplifier (Amp) and computer.

this means incorporating two methods of handling the signal from the photomultiplier tube. These are: (i) for high count rates, a voltage to frequency converter is used to give a train of pulses whose repetition rate is proportional to the signal level; (ii) for low count rates, each individual pulse is counted in a pulse counting mode. The spectra are then collected and stored in a multichannel analyser (MCA) which again is able to cope with the large dynamic range. This MCA can be the same computer as is used for EDX analysis and most commercially-available systems will incorporate some data analysis procedures for quantifying the spectra obtained. The EELS can be used to look at energy losses up to about 5 keV, although at the higher energies the number of counts per channel becomes very small. Generally for light elements the edges of interest occur at energies of no more than 2 keV.

One of the main disadvantages of the serial type of spectrometer is that since each channel of information is collected in turn it has an effective detection efficiency of < 0.05 per cent (i.e. each individual channel is collecting for a time which is the total spectrum collection time divided by the number of channels used for the whole spectrum). However in the new types of EELS with parallel detectors (Ahn and Krivanek, 1986) the spectrum can be collected simultaneously over 1024 channels thus greatly improving their overall efficiency. In these detectors the slit, scintillator and photomultiplier are replaced by a solid-state detector and three quadrupole lenses which are used to increase the dispersion of the magnetic spectrometer (Fig. 4.2B). The solid-state detector consists of a single YAG scintillator and a fibre-optically coupled linear photodiode array. These types of parallel detector have been designed to cope with the large dynamic ranges mentioned above and because the whole spectrum is collected at once, the efficiency is greatly improved to > 10 per cent even though the detection efficiency of the solid-state detector is somewhat lower than for the photomultiplier system. This leads to added advantages such as reduction of radiation damage to beam-sensitive samples, fewer problems with carbon contamination and an improvement of the efficiency for elemental mapping using EELS.

4.2. Spectrometer parameters

When collecting a spectrum the experimental parameters to consider with respect to the spectrometer are the spectrometer entrance aperture, the setting of the slit width (for a serial detector), the energy range to be collected and the parameters to be set on the MCA. As shall be seen in the next section, the entrance aperture plays two roles depending upon the coupling of the microscope to the spectrometer. These are either as an area-defining aperture for selecting the area of the sample being analysed or as an angle-defining aperture for selecting the collection angle to the spectrometer. The rationale for choosing the aperture is described more fully below in Section 4.3.

Sensible selection of the slit width is essential because of its effects both on resolution and on the number of counts in the spectrum. It is important to consider the resolution necessary for the particular application as counts are lost by setting the slit width too narrow (e.g. Fig. 4.3). It is also necessary to consider the resolution of the electron gun being used in the microscope (see Section 4.5), since there is no point in setting the resolution of the spectrometer to less than this as signal is then being discarded unnecessarily. Therefore when setting the slit width, it should always be a compromise between the resolution required and the signal-to-noise ratio. Figure 4.3 shows the effect on the zero-loss peak as the slit is gradually closed. Some spectrometers give a direct readout of the slit width (in micrometers) allowing conditions to be reproduced easily, whereas in others it can only be measured by determining the resolution from the resultant spectrum.

The energy range over which the spectrum is collected is obviously determined by the elemental composition of the specimen. For light elements, this will generally be up to 1 or 2 keV since all the characteristic K edges for elements with $Z < 11$ fall in this range (see Appendix 3). However if higher energy losses are of interest most spectrometers have an energy-offset selector allowing the spectrum to be offset up to 5 keV thus still allowing the range 5–6 keV, for example, to be collected at 1 eV per channel.

Finally, the parameters usually governed by the MCA or computer must be set up before collection of the spectrum can commence. For the parallel spectrometer this involves only the total collection time for the spectrum and the channel positions of any gain changes. However in the serial case these are the dwell time per channel, the number of sweeps of the spectrometer, whether the scan is to be

Electron energy-loss spectrometry 39

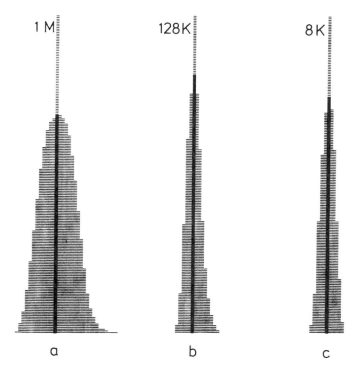

Fig. 4.3. The zero-loss peak collected under identical conditions for three different slit width settings. (a) Wide slit; (b) intermediate; (c) narrow slit. Notice that the full-scale number of counts in each peak is dramatically different.

collected in forward or reverse mode and the positions of any gain changes. The total collection time for the spectrum may be limited by contamination problems, beam damage or specimen drift and these limitations are discussed more fully in Chapter 5. However the time must obviously be maximized in order to improve the statistics of the spectrum. A reverse sweep (i.e. the spectrometer is scanned from higher energy losses down to the zero-loss peak) of the spectrum is recommended if the low-loss region is of interest since the afterglow on the scintillator owing to the high signal intensity of the zero-loss peak may affect the shape in the region close to the zero-loss peak.

4.3. Coupling of the spectrometer to the microscope

There are two optical arrangements that can be used and these are known, rather confusingly, as *image* and *diffraction coupling*. It is important that these terms are fully understood as they allow the setting of different collection angles into the spectrometer. The collection angle, β, is a very important parameter and Fig 4.4 illustrates its simplest definition. Electrons undergoing any type of scattering

40 *Light element analysis in the T.E.M.*

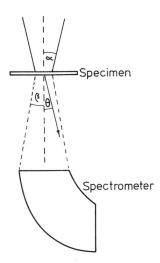

Fig. 4.4. A ray diagram illustrating the definition of the collection angle β of a spectrometer. An incident beam of electrons of convergence angle α will be scattered through an angle θ on passing through the sample.

within the sample will undergo an angular deflection as well as possibly some sort of energy loss. Therefore the emergent electron beam from the sample has an angular distribution as well as an energy distribution. However as seen in the diagram the spectrometer only accepts electrons within the semi-angle β and this can be varied depending upon the coupling of the microscope to the spectrometer. It is important to know the value of the collection angle accurately, as will be seen in Chapter 5 when quantification is discussed. However in practice, it is also important to consider the semi-angle β when collecting the spectrum in order to optimize conditions. This is necessary since the signal-to-noise ratio (SNR), given by P/B where P is the area under the edge and B is the underlying background, is found to vary with β. Joy and Maher (1978) examined this for a range of elements and found that the best SNR for a particular element was found to be for a collection angle of the order of θ_E where

$$\theta_E = E_K/2E_0$$

where E_K is the edge energy and E_0 is the energy of the incident beam. The two types of coupling are now described in turn.

4.3.1. Image coupling

Here the microscope is set up so that a **diffraction pattern is seen on the microscope viewing screen** and therefore there is an image of the specimen at the spectrometer object plane (this is usually the projector lens crossover which approximates to a fixed plane). This is illustrated in Fig. 4.5(a) and it can be seen that the spectrometer half angle of acceptance is $\phi = r/U$ where r is the radius of the entrance aperture and U is its distance from the image. The half angle subtended at the specimen, β, is therefore r/L where L is the camera length. Therefore the collection angle to the spectrometer in this case is determined by both the camera length of the diffraction pattern and the choice of entrance aperture. The area of the sample being analysed is defined by the location of the diffraction aperture.

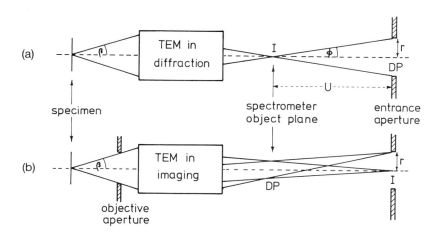

Fig. 4.5. (a) Image coupling of the spectrometer to the microscope: the microscope is set up with a diffraction pattern (DP) on the screen so that an image I is formed at the spectrometer object plane. The spectrometer collection angle β, is then defined by r/L where r is the spectrometer entrance aperture radius and L is the camera length. (b) Diffraction coupling of the spectrometer to the microscope: a normal TEM image (I) is seen on the screen therefore a diffraction pattern (DP) is found at the spectrometer object plane. The spectrometer collection angle β is limited by the objective aperture.

4.3.2. Diffraction coupling

In this case, a diffraction pattern occurs at the spectrometer object plane with the **image of the sample being displayed on the microscope viewing screen** as shown in Figure 4.5(b). Here the spectrometer acceptance angle is determined by the choice of objective aperture and the spectrometer entrance aperture defines the region of the sample from which the analysed electrons are coming. It is important therefore to calibrate the objective apertures so that the corresponding acceptance angle is accurately known for each size of aperture. In order to do this, a diffraction pattern should be set up for a known material (e.g. Al or Au) and photographed so that the pattern can be identified and the camera length can be accurately determined. The same pattern should then be recorded under the same conditions but with each different objective aperture in turn centred around the zero-order (central) beam.

Then by measuring the diameter of the objective aperture d on the negative and using the pre-determined camera length L, the collection angle β can be calculated by geometry as shown in Fig. 4.6.

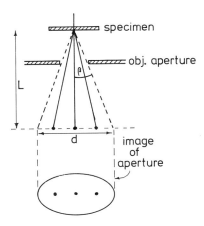

Fig. 4.6. Determination of the collection angle β for diffraction coupling of the spectrometer to the microscope. For a camera length L the image of the objective aperture has a diameter of d (mm) allowing β to be calculated by simple geometry.

4.4. Microscope and sample parameters

With respect to the microscope, the remaining parameter to consider is the operating voltage. Although the ionization cross-sections do vary slightly with kilovoltage, (see Section 5.1.3), generally the overriding factor is that at the higher voltages less multiple scattering occurs since the plasmon mean free path is increased. Therefore it is usual to work at the highest kilovoltages available since one of the main limitations of the technique is preparing samples thin enough. It is important to maintain both sample and microscope contamination-free since the presence of carbon can be a problem because its K edge may overlap with other edges of interest as well as leading to an increased background under the edges. Also when analysing crystalline materials, care must be taken to ensure that the sample is not in a strongly diffracting orientation as this may lead to inaccurate determinations of elemental concentrations (Zaluzec et al., 1980).

4.5. Calibration of the spectrum

An estimate of the calibration is easily obtained since the spectrometer is set to collect over a specified energy range for a certain number of channels. To calibrate the system accurately, two well-defined points are required. One obvious calibration point is the position of the zero-loss peak but for the second a well-defined edge is also required. The most convenient specimen to use for calibration purposes is a thin film of carbon whose edge, as seen in Fig. 4.7, occurs at 284 eV. In an EEL spectrum, even the sharpest edges are spread over several channels so the convention is to use the position midway up the slope of the edge for the edge energy. Calibration may need to be carried out several times during a session since, whilst collecting spectra, the scan coils offset position may need to be changed periodically. It may be possible to recalibrate using a well-defined edge of interest in the specimen, otherwise, ideally, a multi-specimen holder is required in order to have a thin C film available at all times.

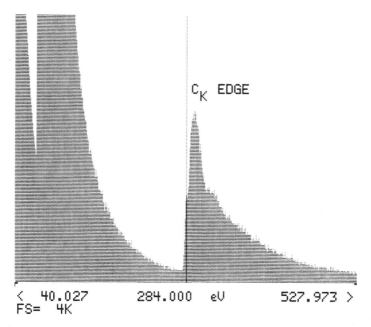

Fig. 4.7. A carbon K edge collected from a thin carbon film showing the position of the edge at 284 eV.

Examination of the C edge also allows the 'jump-ratio' to be determined. This is defined as the height of the maximum point in the edge divided by the height of the background channel immediately preceding the edge. Ideally for C if the spectrometer is well set up the 'jump-ratio' should be in the range 7–10. If this is not the case, it indicates an increased background which should be further investigated (Section 5.3).

4.6. Qualitative interpretation of an EEL spectrum

A typical spectrum can be divided into three distinct regions and the energy-loss processes involved in each region will now be described.

4.6.1. The zero-loss peak

This peak should, in principle, consist of electrons which have lost no energy at all in passing through the sample. However, because the spectrometer has limited resolution, included in the 'zero-loss' peak will be electrons which have lost small amounts of energy for instance by phonon scattering in which the losses involved are typically of the order of 0.1 eV. The electrons contributing to the 'zero-loss peak' can be divided into two groups, the *unscattered* and the *elastically scattered*. Elastic scattering, which effectively involves no energy loss, can occur over a wide range of angles and this may mean that not all elastically-scattered electrons can be collected. It may also affect the inelastic signal since the inelastic electrons (i.e. those which have suffered detectable energy loss) may also undergo an elastic scattering event.

The width of the zero loss signal, usually defined in terms of its FWHM, also gives a measure of the resolution of the spectrometer. Typical values for spectrometer resolution are fractions of an electron volt (for both parallel and serial types of spectrometer) so that usually the limiting factor is the resolution imposed by the electron gun. For example, for a thermionic tungsten filament the energy spread of the electrons is 1–2 eV, for a lanthanum hexaboride the value is <1 eV and for a field-emission gun this value is only about 0.2 eV. Therefore, when setting the resolution of the spectrometer it is important to remember this limitation as was discussed in Section 4.2.

4.6.2. The low-loss region

In this region the intensity of signal obtained is comparable to that of the zero-loss peak, and the shape of the spectrum is determined by energy losses owing to interaction of the fast electrons with the valence or conduction band electrons. The easiest example to consider is the case of metals and alloys where the valence electrons can be considered to form an 'electron gas'. A fast electron passing through the sample may interact with the 'gas', setting up a collective excitation known as a 'plasmon', and losing a small amount of energy in the process. The typical energy loss involved in plasmon excitations is of the order of 15 eV. The mean free path for this excitation is ~ 100nm and therefore in a TEM sample most electrons will undergo such an interaction. Prominent peaks resulting from the excitation of one, two or three plasmons in aluminium can be seen in Fig. 4.8.

For other materials similar interactions often occur but are usually owing to excitation of electrons in various bound states and are less easily definable (Hren *et al.*, 1979). For example in biological samples, fast electrons can interact with the electrons in molecular orbitals giving rise to peaks in similar positions in the spectrum.

Electron energy-loss spectrometry 45

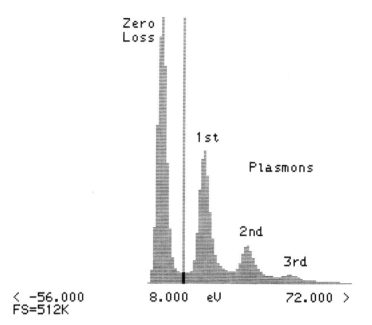

Fig. 4.8. The low-loss region from a spectrum collected from aluminium showing the plasmon peaks found at 15, 30 and 45 eV.

4.6.3. Higher-energy losses

This is where the best analytical information can be found since above 50 eV the ionization edges characteristic of each atom start to appear, sitting on an exponentially falling background. Figure 4.9 shows the K and L edges from aluminium. At high-energy losses, the signal is much reduced compared with the zero-loss region and it is generally necessary to use a gain change to magnify the signal recorded at these energy losses. The background is owing to the tails of features at lower energies and multiple plasmon losses, where an electron has interacted to form a plasmon several times in its passage through the specimen. The characteristic edges correspond to the ionization of the inner shell electrons and correspond to the K, L and M X-ray peaks. For light-element analysis, the edges will generally be K edges. The reason EELS is particularly good for light elements is because the ionization cross-section, σ, increases as Z decreases (see Chapter 1). However it is worth noting that the mean free path for an ionization event to occur is ~ 1 μm, therefore in a 100 nm thick TEM sample, relatively few electrons will cause inner-shell ionizations compared with the number undergoing plasmon excitations. This is why the characteristic edge signal is much lower than the zero-loss peak and the low loss region.

46 *Light elements analysis in T.E.M.*

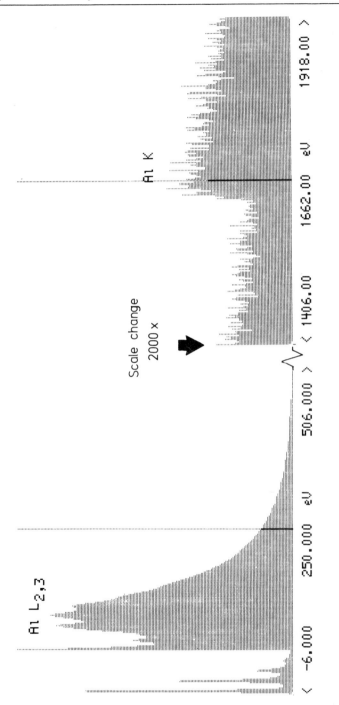

Fig. 4.9. A spectrum collected from aluminium showing the very weak Al_K edge and the much more intense $Al_{L_{2,3}}$ edges.

The use of characteristic edges for qualitative analysis is simply a matter of comparing the edge energy and shape with tabulations of energy (see Appendix 3 or data supplied by spectrometer manufacturers) or compilations of typical spectra (Ahn and Krivanek, 1983). Quantitative interpretation of spectra is considered in Chapter 5.

Quantification of EELS data

This chapter examines the quantification procedure for the ionization edges and looks at the accuracy and detection limits attainable with this technique as well as at some of its limitations. One of the major advantages of EELS is that other types of information, not always analytical, are obtainable using the spectrometer and these are briefly discussed in the last section.

5.1. Quantitative procedure

Figure 5.1. shows a typical spectrum highlighting the features of interest in the quantification procedure. Most commercially-available computers used to collect the spectrum will incorporate a program for data analysis based on the procedure now outlined. One of the first points to notice is the presence of a dark current (this is electronic noise in the signal processing system) in both the analogue and pulse-counting regions of the spectrum. This essentially-constant background should be subtracted from the spectrum before starting the quantification and generally a few channels before the zero-loss peak are used to measure the dark current for the analogue part of the spectrum. For the pulse-counting region, care has to be taken that at the high-energy end it is the true dark current which is being measured and not a contribution from tails of features at lower energy losses. It is also important to measure the gain change, ΔG, in going from the analogue to pulse-counting region of the spectrum.

In the same way as for X-ray analysis, the background must first be removed before summing the counts under the edge. However, in the case of an EEL spectrum the background is not as straightforward since it is of an exponential form and the edge shapes themselves are not as well defined as the Gaussian shape of ideal X-ray peaks. Generally, a region of interest before the edge is used to fit an exponential of the form $I = AE^{-r}$ which can then be extrapolated and subtracted from the edge, where A and r are constants and I is the intensity at the energy E. This is explained in more detail in Section 5.1.1 below. The number of atoms/cm^2, N of a particular element in the sample can then be determined using the following formula:

$$N = \cdot \frac{I_K (\beta, \Delta)}{I \cdot \sigma_K (\beta, \Delta)} \quad (5.1)$$

where $I_K (\beta, \Delta)$ is the intensity of the K edge in the region of width Δ from E_K to $E_K + \Delta$ for a spectrometer collection angle β, $\sigma_K (\beta, \Delta)$ is the partial ionization cross-section for the K edge for electrons scattered within an angle β and within

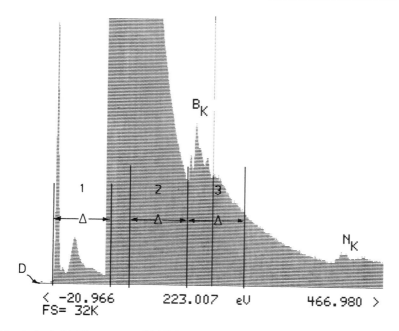

Fig. 5.1. A typical EEL spectrum highlighting the regions of interest for the quantification procedure: (1) is the integral of the zero-loss peak over a width Δ, i.e. $I_1(\beta, \Delta)$; (2) is a region of width Δ for fitting the background; and (3) is the edge integral for the width Δ, i.e. $I_K(\beta, \Delta)$. D represents the dark current for the analogue part of the spectrum.

the energy range Δ and I is the intensity of the incident electron beam. These calculations assume that by using intensity ratios the effect of elastic scattering on the inelastically-scattered electrons will be cancelled out. This is reasonable provided that there are no strong diffraction effects and in practice it is therefore important to tilt the sample away from a strong two-beam orientation. Also, only a single scattering process is considered whereas in fact all processes giving rise to the shape of the low-loss region re-occur at every position on the ionization edge. Provided the sample thickness is less than the plasmon mean free path, λ, a first-order correction is made for this by replacing I with $I_1(\beta, \Delta)$ which is the signal received by the spectrometer under the zero-loss peak for a region of width Δ and collection angle β. Therefore eqn (5.1) now becomes:

$$N = \frac{I_K(\beta, \Delta)}{I_1(\beta, \Delta) \cdot \sigma_K(\beta, \Delta)}$$

In particular, the atomic ratio of two elements A and B may be of interest and it can be seen that the equation then simplifies to:

$$\frac{N_A}{N_B} = \frac{I_{KA}(\beta, \Delta) \cdot \sigma_{KB}(\beta, \Delta)}{I_{KB}(\beta, \Delta) \cdot \sigma_{KA}(\beta, \Delta)}$$

Therefore, the information required is simply the two edge integrals and the ratio of the two partial cross-sections. Practically, this removes the necessity to scan the spectrometer across the intense zero-loss peak and low-loss region which may damage the scintillator. The cross-sections may either be calculated or, if a suitable material of known composition can be obtained containing the two elements, the ratio of their cross-sections may be determined experimentally. In the sections below, we examine in a little more detail the different factors involved in the quantification procedure.

5.1.1. Background fitting

Since the background arises from a complex range of interactions such as multiple plasmon losses and tails of ionization edges at lower energies, it cannot easily be modelled from first principles. As stated above it has been found empirically that an exponential of the form AE^{-r} can be used to give a good fit in most cases. Therefore a region is selected just prior to the edge, the above function is fitted and then extrapolated beneath the edge. The width of this fitting region for the background should be the same as Δ in order to avoid statistical errors in the extrapolation. However the width may be limited by the presence of other edges close by. Also the final channel of this fitting region must be before the start of the ionization edge of interest. To determine whether the fit is good is not easy: the only simple check is to examine the fit at higher energies where it should asymptotically approach the measured data but never actually cross over it (Fig. 5.2). If you are not happy with the fit there is little that can be done except to change the position or the width of the fitting region prior to the edge, and this is not always possible if there are other edges close by. This has led some workers (e.g. Bentley *et al.*, 1981) to propose an alternative method for use when this empirical fit has not been satisfactory. In particular they found background fitting difficult in the presence of M edges and instead tried a polynomial fit to the logarithms of the electron intensity and energy-loss values, generally finding no improvement to the fit in going beyond the second order. The option of using this method is now also available on some of the commercial software but extra care should be taken with this method particularly if extrapolating beyond the edge.

5.1.2. The edges

Once the background has been fitted and extrapolated, the edge integral may be determined by subtracting the background counts in each channel and summing the counts in the region of width Δ starting at the ionization edge energy E_K. Joy and Maher (1978) found the optimum value of Δ to be $\sim 0.15 \, E_K$ to give the best P/B since if Δ is too small, errors can arise from near-edge structure. The structure of the edge depends upon the specific ionization event occurring and for K edges this is a saw-tooth shape whereas L and M edges give a delayed maximum

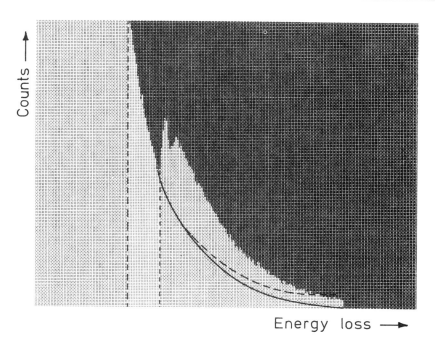

Fig. 5.2. A spectrum illustrating a good fit (———) to the background and a bad fit (- - - - -).

and have a much broader shape owing to the electron-shell structure, as illustrated by the K and L edges of aluminium shown in Fig. 4.9.

5.1.3. Ionization cross-sections

In order to quantify EELS spectra, values are required for the appropriate partial cross-sections as described above in Section 5.1. Calculated values are usually used but if suitable standards are available then experimental values may be obtained for the ratio of two elements. This may be particularly useful for any M edges of interest since at present no theoretical values are available for M edges. However, generally it is difficult to find homogeneous standards of known composition which are also thin enough for EELS analysis.

Theoretically there are two models in use for calculating the values of the partial ionization cross-sections:

1. Egerton (1979) has used a hydrogenic model to calculate the values for the partial ionization cross-sections. Programs containing these calculations are **SIGMAK** (Egerton, 1981a) and SIGMAL (Egerton, 1981b) for the K and L edges respectively, and these can be implemented on any small computer. In fact most commercial software for EELS analysis will have these programs already incorporated. The

program then usually only requires the following values in order to compute the appropriate partial cross-section: β, Δ, the microscope kilovoltage used, the atomic number Z of the element of interest and the type of edge, i.e. K or L.

2. Leapman *et al.* (1980) base their calculations on a more realistic solution of the Schrödinger equation for the isolated atom. However this involves a larger amount of computing to generate some of the values needed.

Both models suffer from the limitation that they only consider transitions to unbound electron states of the free atom so do not give any chemical-shift information. Also, it is important to remember that these are *partial* cross-sections so the value of β should be known as accurately as possible when calculating the appropriate σ (β, Δ). The partial cross-section for a particular element will increase for increasing β but the background underlying the edge also varies with β and there will therefore be an optimum value of β as discussed in Section 4.3. The partial cross-section do also vary with the energy of the electrons used by generally the overriding consideration is that with higher energy electrons multiple scattering will be less of a problem.

5.1.4. Deconvolution

Spectrum deconvolution is often desirable since the effects of specimen thickness are frequently the limiting factor in EELS analysis. The aim of deconvolution is to allow *the spectrum which would have been formed if only single scattering had occurred* to be recovered from the *experimental spectrum in which multiple scattering has actually occurred*. This removes the effects of sample thickness and gives an enhanced edge to background ratio. It is quite a complicated process involving Fourier transforms and although at present no software is available commercially for doing this, some manufacturers are intending to remedy this within the near future. A comparison of the different techniques available for deconvolution is given by Schattsneider and Solkner (1984).

5.1.5. Summary of the quantification procedure

A typical procedure which should be carried out in order to quantify an EEL spectrum is now summarized:

1. Select the parameters required for the quantification, i.e. kV, β, Δ.
2. Subtract dark current from spectrum.
3. Measure the gain change ΔG in going from the analogue to pulse-counting region of the spectrum.
4. Sum counts over the energy range Δ from the zero-loss peak.
5. Specify the position of the edge and set a window marking the region for modelling the background, fit Ae^{-r} and extrapolate.
6. Check the extrapolated fit and, if it is satisfactory, strip the extrapolated region from under the edge: if it is not, return to step 5 and change the fitting region.

7. Sum the counts in the edge region, and if necessary correct the counts for the gain change ($GF = 1/\Delta G$).
8. Calculate the partial cross-section for the element of atomic number Z for the chosen ionization event, i.e. for a K or L edge.
9. Calculate the number of atoms present per square centimetre of that particular element.
10. Return to step 5 and repeat for any other edges of interest.

5.2. Detection limits and accuracy

The detectability of an edge or peak has been discussed in Chapter 3 and here mention will briefly be made of some of the parameters which will affect the statistics of the number of counts in the edge.

Generally, the number of counts can be maximized by increasing the counting time per channel and using the brightest type of electron gun within the limitations of radiation damage to the specimen and specimen drift. Also the values of β, Δ and the kilovoltage used can all be optimized to give the best peak-to-background ratio for the element of interest.

The main difference between the X-ray and EELS cases when considering detection limits is the much higher background for EELS and the fact that the background under the edge has to be extrapolated from the low energy side of the edge. This extrapolation procedure can lead to large errors which are extremely difficult to assess and may make the detection limits somewhat worse than might be expected from an initial examination of the statistics involved.

The accuracy of the technique is obviously affected by the number of counts in the edge and also by the accuracy of the partial cross-sections used. For the calculated cross-sections, β must be known as accurately as possible for the microscope conditions used, in order to avoid significant errors in σ_K or σ_L. Discrepancies have also been reported when examining compounds containing both K and L edges of interest (Sklad et al., 1984) indicating there is not always good agreement between the calculated ionization cross-section values for the K and L edges.

5.3. Limitations of the technique

Undoubtedly the main limitation of EELS is the need for a very small sample thickness, which makes sample preparation for this technique very important. The underlying assumption in simple EELS analysis is that each electron has only undergone one scattering event, whereas in practice the edge consists of a convolution of multiple events. As the sample thickness increases the edge signal rises to a maximum and then falls away whilst the background rises steadily. It is generally recommended that the sample thickness should be less than the total mean free path for inelastic scattering, i.e. for most metals the thickness should be <100 nm and for biological sample the ideal thickness would be 30–50 nm. An easy way to monitor this is to compare the intensity of the first plasmon peak with that of the zero loss peak, since this can be used to give a value for thickness (Section 5.4.1). The edge-shape changes as the electrons are multiply scattered and adding this to the effect of the increasing background leads to a reduction in the visibility of the edge (see Fig. 5.3). In principle, deconvolution of the spectrum to give the single scattering profile, discussed above in Section 5.1.4, will therefore be an important step forward, when it becomes routinely available as part of the commercial software, in solving some of the problems of sample thickness.

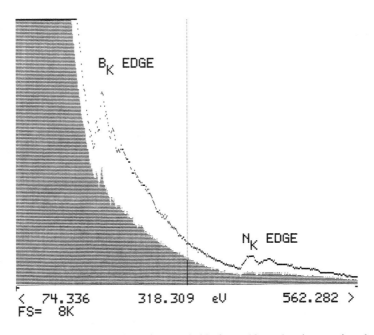

Fig. 5.3. Two spectra collected from boron nitride for a thin region (- - - - - -) and a thicker region (————) of the sample.

It is important to keep both specimen and microscope as free from contamination as possible. The build-up of carbon on the sample has two effects:

(i) the presence of the C edge may overlap with other edges of interest and also gives an increased background;
(ii) it increases the thickness of the sample thus bringing in the problems of multiple scattering discussed above.

It is therefore important to minimize contamination wherever possible.

The spectrometer itself should also be checked carefully to ensure that no spurious electron scattering is occurring. This is easily done by collecting a spectrum with no sample present; if a clean spectrum is obtained then no scattering is occurring. However if spurious signals are seen, then further action may be required and the reader is referred to Joy and Maher (1980) who discuss this in more detail.

Several problems can arise with a serial spectrometer which would not occur to such an extent in the parallel type. In particular these are the effects of specimen drift during the acquisition of the spectrum and the effects of radiation damage on beam-sensitive materials. In a serial spectrometer these effects can manifest themselves in two main ways:

(i) owing to the longer collection times required compared to parallel detectors both effects will be much greater; and
(ii) because of the serial collection of the spectrum an edge at a higher energy may have been collected from a different part of the sample or from a more radiation damaged region owing to the time difference in collecting the two different edges.

Also, of course, the longer collection times for a serial spectrometer will give a greater opportunity for contamination to build up. Therefore the advantages of the parallel spectrometer are quite significant. Finally, quantification will be limited when edges overlap since the background cannot be fitted and extrapolated and therefore the number of counts in the ionization edge cannot be determined.

5.4. Other information obtainable

An EEL spectrometer can also be used to give other types of information which can conveniently be divided into three groups: Information from the low-loss region, chemical information from edge structure and energy filtered images. Each of these three groups will now be briefly outlined.

5.4.1. Low-loss spectrum

As mentioned previously in this chapter and in Chapter 3, the low-loss region of the spectrum gives a relatively easy and convenient way of determining the sample thickness. The production of plasmons has been described in Section 4.5 and the probability $P(m)$ of an electron exciting m plasmons in travelling through a sample of thickness t is given by the Poisson distribution, i.e.

$$P(m) = \frac{1}{m!} \left(\frac{t}{\lambda_P}\right)^m \exp(-t/\lambda_P)$$

Therefore the ratio of the probabilities of exciting the first plasmon and no plasmons is

$$\frac{P(1)}{P(0)} = \frac{t}{\lambda_P}$$

and this can be determined experimentally by taking the ratio of the first plasmon peak to the zero-loss peak. In order to determine thickness the plasmon mean free path, λ_P, must be known for the material of interest. Some values are tabulated in the literature, e.g. Raether (1965 and 1980) but generally it is necessary to use a sample of the same composition and of known thickness to determine the mean free path for the conditions being used. Table 5.1 gives some typical values that have been determined in our laboratory for λ_P but they should be used with caution since λ_P depends on the collection angle, β and the energy of the primary electrons. However once λ_P has been determined, for a particular set of conditions, the 'plasmon ratio' is a quick and fairly easy measurement to make.

Table 5.1. λ_P for $\beta = 30$ mrads

Element	Kilovoltage		
	120kV	100kV	80kV
Al	1284	1257	1122
Si	1350	1187	1158
Mo	2141	2010	1827

Other relationships have also been derived, one of the most widely used being the equation:

$$t/\lambda_P = \log_e (I_T/I_0)$$

Where I_T is the total spectrum intensity measured by integrating the spectrum over an energy window of 200 eV or more.

Plasmons have also been used to give chemical information in certain alloys where the addition of elements has been found to shift the plasmon peak position. By careful monitoring of the peak position with the addition of the element, the system can then be calibrated to give the concentration of that element as a function of shift in electron-volts. This was some of the earliest work done in EELS and was mainly carried out for aluminium alloys (Williams and Edington 1976) where the plasmon peak is well defined, e.g. the variation of elements such as Mg, Cu and Li in Al alloys. However, it is worth noting that one requirement is fairly good resolution since the shifts measured may only be fractions of an electron-volt.

5.4.2. Structure in the ionization edges

Structure may be visible both at the edge itself and extending several hundred electron-volts away from the edge. There are two types of fine structure, illustrated diagramatically in Fig. 5.4:

1. *Energy loss near-edge structure (ELNES)* is found within 20 or 30 eV of the edge itself. The chemical binding state of the element in the specimen affects both the shape of the edge and the actual edge energy. For this technique the spectrometer needs to be set up to give good resolution.

2. *Extended energy-loss fine structure (EXELFS)* is found in the form of weak

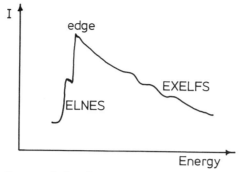

Fig. 5.4 A schematic diagram of the edge structure showing the positions of the energy-loss near-edge structure (ELNES) and the extended energy-loss fine structure (EXELFS).

oscillations sitting on a monotonically decreasing background for several hundred electron-volts beyond the edge. It provides information on the number and type of neighbours surrounding a particular chemical species and is analogous to the older-established technique EXAFS (extended X-ray absorption fine structure).

For both techniques it is necessary to have as good an edge signal as possible and therefore the effects of multiple scattering will be important, making sample thickness a critical parameter. To extract useful information considerable data processing is required in each instance therefore making these techniques only available in a few specialist laboratories at present (Johnson *et al.*, 1981; Krivanek *et al.*, 1982; Colliex *et al.*, 1985).

5.4.3. Elemental mapping and energy filtering

In Chapter 2, the technique of X-ray mapping was briefly described and the same principle may be applied to EELS. By stopping the spectrometer scan and allowing only electrons of a particular energy to reach the detector, a map can be collected from a scanned image or diffraction pattern showing the intensity variation throughout the image for that energy. In this way, by choosing the edge energy for the element of interest a distribution of that element over an area of the sample can be mapped. However it is very important when doing this that a window is set immediately before the edge to collect a map showing the intensity distribution of the background underlying the edge otherwise spurious effects owing to thickness variations may be seen. Therefore once the background and edge maps have been collected the first may be substracted from the second to show the true elemental variation. Figure 5.5 is an example of such an elemental map collected by EELS and it shows the variation of B in TiB_2 particles on a thin C film. The maps from the B edge and for the background preceding the edge show clearly the importance of background subtraction.

An advantage of EELS mapping over X-ray mapping is that the spectrometer can be used to map not only with ionization edges but with other electrons of a

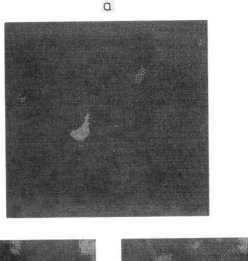
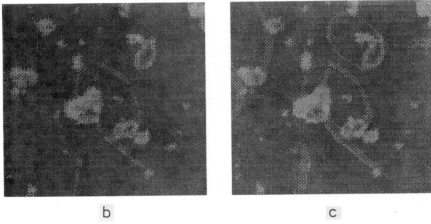

Fig. 5.5. (a) An energy-loss map collected for the B_K edge for a TiB_2 particle on a thin film, after background subtraction, i.e. map (b) minus map (c). (b) This is the actual map collected from the B edge. (c) The map collected from the background immediately preceding the B edge.

particular energy. This means therefore that an image can be collected using, for example, the electrons from the plasmon peaks or from the zero loss peak. Using only the zero-loss electrons to form an image means that no inelastically-scattered electrons contribute, giving a normal bright field image collected with a purely elastic signal. The spectrometer can be used in this way to energy filter not only images but also diffraction patterns, enhancing both contrast and resolution.

Finally, in biological samples this is used to give Z contrast in the image by dividing the elastic image by the inelastic signal. This is not really applicable to images of crystalline materials since the diffraction contrast interferes with the low intensity of the Z contrast images. Therefore this technique has mainly been applied to heavy molecule tagging in organic systems.

WEDX versus EELS: Comparison of the two techniques

In this chapter a comparison is made of the two techniques with examples illustrating some of their differences. Ideally all analytical microscopes would have attached to them an EELS and a windowless detector. However often financial considerations determine that only one of the two is available so below we list some of the factors to be considered if such a choice has to be made.

6.1. Range of elements detectable

At present windowless X-ray analysis allows elements down to B to be detected with a TEM although reports have been made of Be detection using an SEM (Statham, 1986). However EELS has been used to detect Li (Sainfort and Guyot, 1985), He and the presence of hydrides (Thomas, 1981). Detection of these elements does require good signal and resolution since the edges of interest are superimposed on the fairly high intensity of the low-loss region and will have to be distinguished from any structure present there. This makes any sort of quantification difficult because of problems in modelling and removing the background from under the edge.

It is also important to consider the other elements of interest in the sample since very often the problem to be solved involves quantifying the elemental variation in a particular region. Here X-ray analysis definitely has the advantage since it is able to detect much more easily the higher Z elements in the material. Two examples from our laboratory can be used to illustrate this point. In the first, the distribution of Li in rapidly solidified Al/Li powder was being examined. The detection of Li in Al is a well-known problem for EELS because of the overlap of the third Al plasmon peak with the Li edge and quantitation is also difficult (Chan and Williams, 1985). This was overcome by good sample preparation to obtain sufficiently thin areas for analysis and in the Li rich regions we were able to detect and to attempt to quantify the amount of Li present (Von Bradsky and Ricks, 1987). Figure 6.1 shows a spectrum of such a Li edge next to the $Al_{L_{2,3}}$ edge. However examination of the EEL spectrum at higher energies revealed no further information on any other elements which might have been present in these regions. Only by further analysis using a conventional X-ray detector could the distribution of the higher atomic number elements be determined. A spectrum obtained from such a segregation region for a powder which has been Ni-plated and then ion-beam thinned, (Fig. 6.2) shows that in fact Fe, Cu and Mg are also present.

60 *Light elements analysis in T.E.M.*

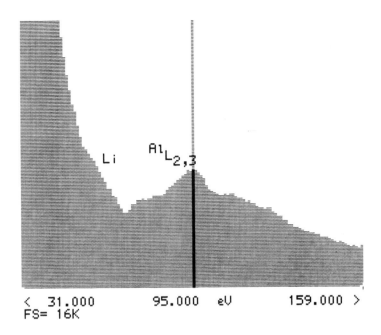

Fig. 6.1. An EEL spectrum from the intercellular region of an Al/Li powder showing the Li_K edge and the $Al_{L_{2,3}}$ edge.

The second problem involved the analytical study of a boron-containing glass obtained during the grain refining of a brass (Gregg *et al.* 1985). The glass was extracted from the brass by carbon replication thus allowing the sample to be analysed in the TEM. The main constituent of the glass is B ($\sim$ 90 per cent) and consequently the B edge is easily detected in the EEL spectrum. However, it is difficult to detect any of the other elements present. In some of the samples where the O content is fairly high, a small O edge is seen and so the B to O ratio can be determined (Fig. 6.3). With the windowless X-ray detector it is more difficult to obtain a good signal at the B peak but the O peak is clearly seen along with the other five or six elements present in the sample in much smaller quantities (Fig. 6.4). For quantification in the X-ray case, there are of course still problems such as the value of the K factor for B (Budd and Goodhew, 1986) and the accuracy of the mass absorption coefficients used. However the X-ray data do appear to be consistent and reproducible when examining a variety of particles with different O to B ratios. Table 6.1 below shows the composition of two different types of B-containing particle analysed in three different regions for each one.

These two examples illustrate clearly that it is advantageous to have both techniques available to be able to obtain the optimum information for solving the problem.

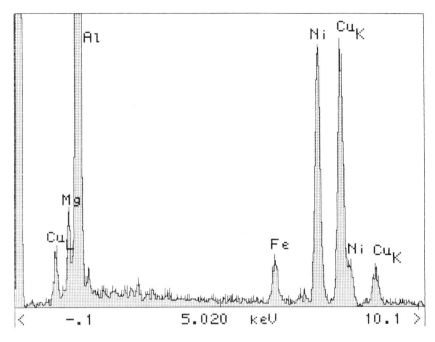

Fig. 6.2. An X-ray spectrum from the intercellular region of an Al/Li powder showing the segregation of Mg, Cu and Fe which could not be seen in the EEL spectrum. The sample was prepared by Ni plating the powder and then ion-beam thinning.

Table 6.1

Sample	Elemental concentrations (weight per cent)					
	B	O	Al	Fe	Ca	Zn
A	85.4	5.9	5.0	0.1	0.5	2.4
	85.4	5.6	6.1	0.1	0.2	2.3
	86.3	3.5	6.6	0.3	0.2	2.8
B	75.5	17.5	1.5	0.3	0.2	0.3
	75.6	15.6	4.4	2.3	0.3	0.6
	73.1	15.8	3.4	2.1	2.1	1.1

6.2. Thickness limitations

As already discussed in Chapters 3 and 5, thickness will be a limiting factor in both techniques, making sample preparation even more important for light-element analysis than it is for conventional analytical TEM work. For EELS, multiple scattering of the electrons causes the edge visibility to decrease as the sample increases in thickness. For X-ray analysis the thickness leads to absorption of the

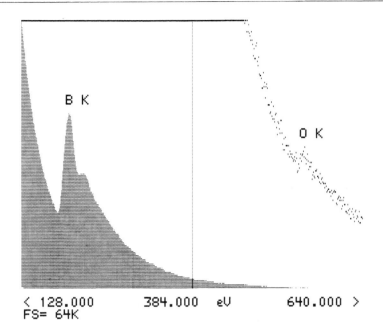

Fig. 6.3. An EEL spectrum from a B-rich glass showing the B_K and O_K edges.

X-rays, again reducing the size of the signal from the light element and although this can be corrected for, there may be large inaccuracies in the mass absorption coefficient data. Therefore from a qualitative point of view, the windowless X-ray detector is probably more useful when thickness is a problem since the presence of a particular peak may at least be determined. From a quantitative point of view, the inaccuracies incurred in allowing for thickness are probably equally great in each case.

6.3. Overlap problems

Owing to the inherently poorer resolution of X-ray detectors, overlap problems are a much greater problem than in EEL spectra. For the metallurgist there will be difficulties when looking at transition metals from the overlap of the L lines with the K lines of any light elements, particularly the O_K line. For example, Fig. 6.5 shows the corresponding X-ray and EEL spectra for a titanium chromium oxide. For the biologist problems may occur with the calcium L peak falling between the carbon and nitrogen K lines as illustrated by Fig. 6.6 which shows the X-ray and EEL spectra collected from a gallstone. Both these examples show the better separation of edges in the EELS case compared with the corresponding X-ray spectrum. However, quantification will become difficult in both cases since for the EELS it will not generally be possible to fit the background owing to the close

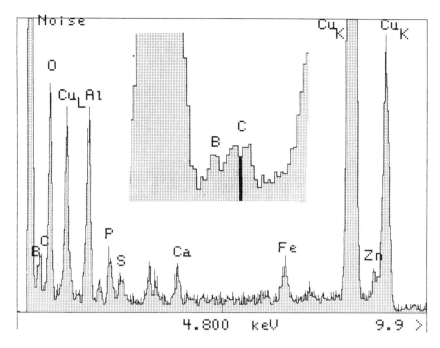

Fig. 6.4. An X-ray spectrum from a B-rich-containing glass showing a range of elements present in smaller quantities. The inset shows an enlargement of the low-energy end of the spectrum with the B and C peaks.

proximity of the other edges. For the X-ray case, good spectrum deconvolution procedures will be required to separate the overlapping peaks which will limit the accuracy and where the overlap is particularly severe this will probably not even be possible.

6.4. Contamination

Again the presence of carbon contamination will be a problem for both techniques owing to overlap with other peaks or edges of interest. With EELS there will also be a higher background above the C edge and the increase in sample thickness will give more multiple scattering. For the X-ray case an extra layer of C will give increased absorption of the low energy X-rays. A particular example of interest for the X-ray case occurs when looking for B since the presence of a large C peak may mask the small B peak next to it. Figure 3.7 (p. 32) shows a spectrum from pure B on a C film and a small C peak from the support film can be seen overlapping with the edge of the B peak. However in this case the information may be recoverable from the spectrum if good peak-fitting routines are available for the data analysis.

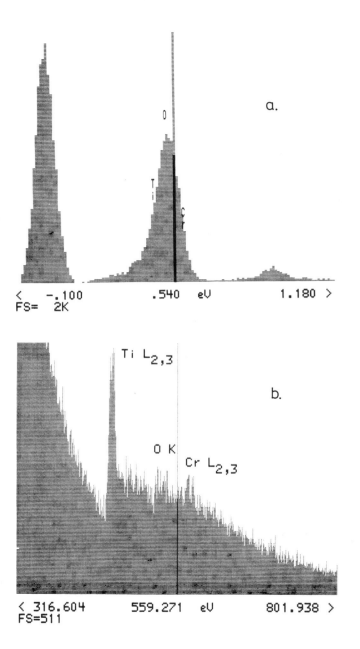

Fig. 6.5. The X-ray (a) and EEL (b) spectra collected from a titanium chromium oxide illustrating the problems of overlap for the metallurgist.

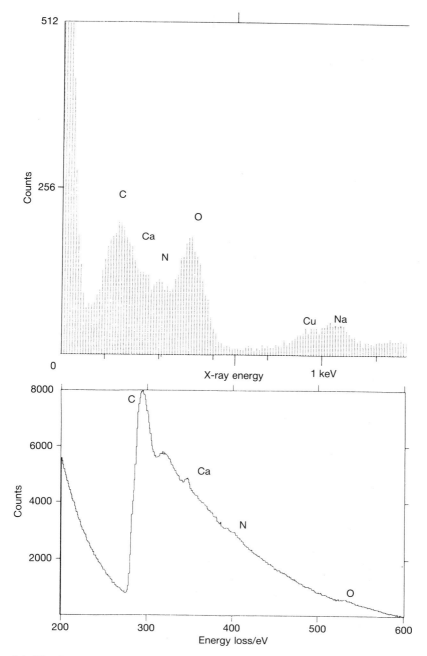

Fig. 6.6. The X-ray (a) and EEL (b) spectra from a gallstone showing the type of overlap problems that may arise for the biologist.

6.5. Other points of comparison

The spatial resolution limit of both techniques is simply that owing to beam-broadening effects. However in reality, as smaller and smaller probe sizes are used for the analysis then the sensitivity will decrease as a result of the very low signals detected in each case. Therefore, in practice, the limit will probably be set by the X-ray count rate or number of counts in the edge necessary to obtain a statistically significant result.

For EELS mapping it is important to have good image analysis routines in order to be able to subtract accurately the background from under the edge, otherwise the map primarily gives information about thickness. For serial spectrometers, since each channel is collected in turn for each element of interest, ideally three channels are required (two to model the background and one for the edge itself). A good count rate is required to obtain good statistics in the image in the time limit set by specimen or spectrometer drift considerations. However the advent of parallel spectrometers will remove this limiting factor and allow the maps to be collected in a much shorter time. X-ray mapping is also limited by count-rate considerations since at low energies, owing to effects such as absorption, the signal may not be very large. X-ray mapping has the added disadvantage that the presence of peak overlaps will be a severe problem since overlapped peaks cannot easily be separated for the purpose of mapping.

With respect to the other information that the two techniques can provide, EELS must surely have the advantage. An X-ray detector may only be used to give an estimate of thickness variations in a sample by one of two methods:

(i) if there is a suitable element with both L and K lines present their ratio can be monitored; or
(ii) calibration of the variation of intensity of a particular peak with thickness can be carried out if a suitable element of constant concentration is present in the material.

In contrast EELS can be used to give a wide range of different types of information. Several useful techniques for analysis as well as thickness determination have been briefly covered in Section 5.4.

A final consideration must be the ease with which each technique can be used. From a user's point of view a windowless X-ray detector is much easier to use than an energy-loss spectrometer particularly because most TEM operators are familiar to some extent with the conventional Be window type of detector. Also, for interpreting the spectra qualitatively X-ray analysis has the advantage since the Gaussian-shaped peaks are more easily identified from a background which is much lower than in the EELS case. An estimate of elemental concentrations may be obtained for materials where no peak overlap occurs by simply examining windows set around the peaks of interest whereas for EELS to obtain any quantitative information the full procedure described in Chapter 5 must be carried out. However with respect to sample position when analysing, EELS has the advantage, since there are no problems owing to shadowing as there are with X-rays. During X-ray analysis the user must take extreme care to ensure that neither the grid bars, nor the specimen holder nor the sample itself are shielding the detector from the X-rays. This problem does not arise during EELS since the analysed electrons remain close to the optical axis of the microscope and thus the signal entering the spectrometer is essentially the same as that seen by the microscopist.

Appendices

Appendix 1. *Some of the more useful low-energy X-ray lines*

Energy (eV)	K	Lα		Lβ	Mα		Mβ
100	110 Be						
200	185 B						
300	282 C	341	Ca	344			
400	392 N	395	Sc	399			
500	523 O	452	Ti	458			
		510	V	519			
600		571	Cr	581			
		636	Mn	647			
700	677 F	704	Fe	717			
800		775	Co	790	833	La	854
900	851 Ne	849	Ni	866	883	Ce	902
		928	Cu	948	929	Pr	950
1000	1041 Na	1009	Zn	1032	978	Nd	997

Appendix 2. *Mass absorption coefficients in cm^2/g for light-element K lines.*
Notes: **Bold** refers to Henke and Ebisu, *Italic* refers to Bracewell and Veigele and other figures refer to Sandborg and Merkle. For other values, particularly for higher-energy lines, see Heinrich for a tabulation or Thin and Leroux for empirical expressions from which the coefficients can be calculated.

Absorber	B	C	N	O	F	Ne	Na
Li	**31 590**	**10 310**	**3812**	**1602**	735	366	197
Be	**60 560**	**22 000**	**8875**	**3922**	1878	963	530
B	3353	37 020	15 810	7416	3677	1947	1093
B	*5420*	*36 900*	*14 800*	*6990*			
C		2194	25 122	11 045			
C	6456	2373	25 490	12 380	6366	3457	1984
C	*6760*	*2380*	*23 400*	*11 900*			
N		3953	1586	17 506			
N	**10 570**	**3903**	**1637**	**17 310**	9161	5069	2970
N	*10 600*	*3920*	*1640*	*19 200*			
O		6518	2614	1186			
O	**16 530**	**6044**	**2527**	**1200**	12 390	6961	4154
O	*16 600*	*6280*	*2690*	*1340*			
F		10 055	4033	1830			
F	**23 880**	**8730**	**3602**	**1688**	868	8647	5216
Ne		14 731	5909	2681			
Ne	**35 760**	**13 570**	**5617**	**2582**	1301	715	6931
Na		20 707	8306	3768			
Na	**46 010**	**18 460**	**7796**	**3645**	1823	984	561
Mg		28 145	11 289	5122			
Mg	**58 170**	**24 900**	**10 950**	**5174**	2615	1411	815
Al		37 197	14 920	6769			
Al	**65 170**	**30 160**	**13 830**	**6715**	3407	1848	1068
Al	*116 000*	*38 600*	*14 700*	*6830*			
Si		48 016	19 260	8738			
Si	**74 180**	**36 980**	**17 690**	**8790**	4543	2484	1442
Si	*141 000*	*47 900*	*18 600*	*8770*			
Ca		20 221	63 285	28 711			
Ca	**13 010**	**6838**	**35 590**	**22 030**	12 370	7163	4359
Fe		41 644	17 446	8219			
Fe	**25 780**	**13 300**	**7121**	**4001**	2328	13 210	8204
Fe	*326 000*	*102 000*	*36 500*	*16 300*			
Ge	45 050	25 520	14 040	7867	4539	2737	1712
Sr	27 340	35 310	23 310	13 760	8210	5025	3189
Mo	4717	32 420	23 220	18 660	11 360	7087	4544
Cd	6567	5793	4152	20 190	14 870	10 440	6819
I	5880	7094	5613	3906	17 240	10 170	8374
La	3826	7894	6410	4690	3277	12 340	7930

W	19 660	18 750	13 880	10 990	7928	5358	3672
W	*562 000*	*191 000*	*74 000*	*38 400*			
Au	8952	15 210	15 440	11 760	9289	6762	4699
Bi	3136	9533	13 810	12 690	9285	7380	5464
U	2247	2317	2417	11 060	9492	8420	6092
Mylar	9615	3516	16 780	8140	8106	4479	2624

Appendix 3. *Some of the low-energy edges (in eV)*

K		L_2		L_3		M_4		M_5
55	Li	74	Al	73		70	Br	69
		100	Si	99		89	Kr	89
111	Be					112	Rb	111
		136	P	135		135	Sr	133
		165	S	164		160	Y	158
188	B					183	Zr	180
		202	Cl	200		208	Nb	205
						230	Mo	227
		247	Ar	245		257	Tc	253
284	C					284	Ru	279
		297	K	294		312	Rh	307
						340	Pd	335
		350	Ca	347		373	Ag	367
399	N	407	Sc	402		411	Cd	404
						451	In	443
		461	Ti	455		494	Sn	485
532	O	520	V	513		537	Sb	528
		584	Cr	575		582	Te	572
						631	I	620
686	F	652	Mn	641		685	Xe	672
		723	Fe	710		740	Cs	726
		794	Co	779		796	Ba	781
						849	La	832
867	Ne	872	Ni	855		902	Ce	884
		951	Cu	931		951	Pr	931

References

Ahn, C.C. and Krivanek, O.L. (1983). *EELS Atlas*. (A joint project of the ASU HREM facility and GATAN).
——, —— (1986). *Proceedings of the 44th Annual Meeting of the Electron Microscopy Society of America*, pp. 618–619, San Francisco Press.
Allen, S.M. and Hall, E.L. (1982). *Phil. Mag.* **A46**, 243–53.
Bentley, J., Lehman, G.L., and Sklad, P.S. (1981). *Analytical Electron Microscopy EMSA*, R.H. Geiss (ed.), San Francisco Press.
Bracewell, B.L. and Veigele, W.J. (1971). Tables of X-ray mass attenuation coefficients for 87 elements at selected wavelengths. *Developments in Applied Spectroscopy* **9**, E.L. Grove and A.J. Perkins (eds.), pp. 357–400, Plenum, New York.
Budd, P.M. and Goodhew, P.J. (1986). *Microbeam Analysis* A.D. Romig and W.G. Chambers, (eds.), San Francisco Press.
Chan, H. and Williams D.B. (1985). *Phil. Mag.*, **B 52** 1019–32.
Colliex, C., Manoubi, T., Gasgnier, M., and Brown, L.M. (1985). *Scanning Electron Microscopy*, **2**, 489–512.
Egerton, R.F. (1979). *Ultramicroscopy*, **4**, 169–79.
——, (1981a). *J. Microscopy*, **123** (3), 333–7.
——, (1981b). *Proceedings of the 39th Annual Meeting of the Electron Microscopy Society of America*, pp. 198–9, San Francisco Press.
Gregg, N.R., Budd, P.M., and Goodhew, P.J. (1985). *Proc. EMAG 85, Inst. Phys. Conf. Ser.* No. 78, pp. 277–80.
Hall, T.A. and Gupta, B.L. (1979). *Introduction to Analytical Electron Microscopy*, J.J. Hren, J.I. Goldstein and D.C. Joy (eds.), pp. 169–197. Plenum Press, New York.
Heinrich, K.L.F. (1966). 'The electron microprobe', T.D. McKinley, K.F.J. Heinrich and D.B. Wittry (eds.), Wiley, New York.
Henke, B.L. and Ebisu, E.S. (1973). *Advances in X-ray Analysis* **17**, 150–213.
Hren, J.J., Goldstein, J.I. and Joy, D.C. (1979). *Introduction to Analytical Electron Microscopy*, Plenum Press, New York.
Johnson, D.E., Csillag, S. and Stern, E.A. (1981). *Scanning Electron Microscopy*, **1**, 105–15.
Joy, D.C. and Maher, D.M. (1978). *Ultramicroscopy*, **3**, 69–74.
—— and —— (1980). *Scanning Electron Microscopy*, **1**, 25–32.
Kelly, P.M., Jostsons, A., Blake, R.G., Napier, J.G. (1975). *Phys. Stat. Sol. (A)*, **31**, 771–80.
Krivanek, O.L., Disko, M.M., Tafto, J., and Spence, J.C.H. (1982). *Ultramicroscopy* **9**, 249–54.
Leapman, R.D., Rez, P., and Mayers, D.F. (1980). *J. Chem. Phys.*, **72**, 1232.

Morgan, A.J. (1985). *X-ray Microanalysis in Electron Microscopy for Biologists*, RMS Handbook 5, Oxford University Press, Oxford.
Raether, H. (1965). *Proceedings of the International Conference on Electron Diffraction and Crystal Defects*.
——— (1980). Excitation of plasmons and interband transitions by electrons. *Springer Tracts in Modern Physics*, 88, Springer Verlag, New York.
Rong, W., Wirmark, G., and Dunlop, G.L. (1984). Analytical electron microscopy – 1984, D.B. Williams and D.C. Joy (eds.), pp. 51-6, San Francisco Press.
Sainfort, P. and Guyot, P. (1985). *Phil. Mag.*, **A51**, 7-20.
Sandborg, A.O. and Merkle, A.B. (1981). *Scanning Electron Microscopy*, 1, 63-70.
Schattsneider, P. and Solkner, G. (1984). *J. Microscopy*, 134 (1), 73-87.
Sklad, P.S., Bentley, J., Angelini, P., and Lehman, G.L. (1984). *Analytical Electron Microscopy*, D.B. Williams and D.C. Joy (eds.), p. 285, San Francisco Press.
Statham, P.J. (1986). *Microbeam Analysis*, A.D. Romig and W.F. Chambers (eds.), pp. 281-4, San Francisco Press.
Thin, T.P. and Leroux, J. (1979). *X-ray Spectrometry*, 8, 85-91.
Thomas, G.J. (1981). *Analytical Electron Microscopy* R.H. Geiss, (ed.), San Francisco Press.
Von Bradsky, G.J. and Ricks, R.A. (1987). Submitted for publication.
Williams, D.B. and Edington, J.W. (1976). *J. Microscopy*, 108 (2), 113-45.
Zaluzec, N.J. (1979). *Introduction to Analytical Electron Microscopy*, J.J. Hren, J.I. Goldstein and D.C. Joy, (eds.) Ch. 4, pp. 121-167. Plenum Press, New York.
———, Hren, J.J. and Carpenter, R.W. (1980). *Proceedings of the 38th Annual Meeting of the Electron Microscopy Society of America*, p. 114.

Index

aberrations, spectrometer 35
absorption 10, 16, 18, 20, 22ff, 26
accelerating voltage 42
accuracy 30
AES 1
analysis checklists 14, 16, 52
Auger electrons 1, 2

background
 in EELS spectra 45, 48, 53
 fitting 50
 subtraction 20, 30, 48, 50, 57, 59
backscattered electrons 10
beam damage 38, 39, 55
beryllium 59
beryllium window 4
boron 6, 11, 12, 21, 24, 31, 57, 60, 63
boron 16, 17, 21, 54
Bremsstrahlung 9, 11

calcium 62
calibration 42
camera length 41
carbides 25, 29
carbon 21, 24, 25, 31, 42, 54, 62
CBED 28, 29, 30
characteristic edges 45, 50
characteristic X-rays 1, 3, 18
collection angle 39–41
contamination
 of detector 15
 of specimen 28, 31, 38, 42, 54, 63
contamination spot method 28
continuum method 20, 27
convergent beam thickness determination 28
count rate 12
coupling 38, 39–42
cross-sections 1, 3, 18, 42, 45

dark current 48, 52
dead layer 7
deconvolution 18, 30, 31, 52, 54
detection efficiency 37
detection limit 30ff, 53
detector-specimen distance 10
diffraction aperture 41
diffraction coupling 41

diffraction effects in EELS 42, 49
drift 53, 55
dynamic range 35

EELS 1
edges 45, 50, 56
EDX 44, 49, 58
electron gun 44, 53
elemental mapping
 EELS 57, 66
 X-ray 66
ELNES 56
energy filtering 57
entrance aperture, spectrometer 38, 41
escape peak 13, 14, 16
excitation efficiency 18
EXAFS 57
EXELFS 56
extraction replicas 25, 60

Fano factor 7
fluorescence 13, 15
fluorescence yield 3, 18

gain 12
 changes 45, 48, 52

helium 59
homogeneity of specimen 23, 24, 31
hydrogen 59

icing 15
image coupling 40
incomplete charge collection 7
inner-shell cross-section 1, 45
ionization cross-sections 48, 51
ionization edges 49–51, 56, 57
ionization energy 3

jump ratio 43

K factor 20, 27, 30
K/L ratio 15, 29

light 10
lithium 59
low energy tailing 7
low-loss region 44, 55

mapping 16, 38, 57, 66
mass absorption coefficient 22, 24, 26, 60, 62
minimum detection limits 30
multi-channel analyser (MCA) 37
multiple scattering 52, 54, 61, 63

nitrogen 21, 62
noise 7, 9, 12, 48

organic materials 58
overlaps 16, 18, 20, 26, 30, 32, 62
oxides 22, 24
oxygen 24, 31, 60

parallel detection (EELS) 4, 35, 55, 66
partial cross-section 51, 52, 53
particles 25, 57, 60
peak to background ratio (P/B) 8, 9, 50, 53
peak overlap 16, 18, 20, 26, 30
phonons 44
photomultiplier 35, 37
plasmon 44
    mean free path 42, 44, 56
    ratio technique 27, 55
precipitates 25
processing time 12
pulse counting 37, 48

qualitative analysis 16, 43
quantitative analysis 26ff, 48ff.

radiation damage 38, 53, 55
ratio method 20, 27
replication 25, 60
resolution 7, 35, 44, 66
reverse EELS collection 39

scintillator 35, 37
serial spectrometer 35-38, 55, 66
SIGMAK, SIGMAL 51
signal/noise ratio (SNR) 40
silicon nitride 21
single scattering 49, 52
slit width 35, 38, 39
sodium 24
specimen drift 53
    homogeneity 23, 24
    preparation 61
    tilt 9, 10, 22
standard peak profiles 20
standards 21, 51
statistics 17, 30, 52
strobe peak 12
sum peak 12, 16
surface layers 33

tailing 7
take off angle 9, 24, 25
thickness 23, 31, 53, 57, 61
    determination 27-30, 55, 66
thin film criterion 20
threshold 12
tilt 9, 10
turret 10

ultra thin window (UTW) 6, 10, 14

voltage-to-frequency converter 37, 48

Wentzel equation 3
windowless EDX 4

X-ray intensity ratio 27

Z contrast 58
zero-loss peak 35, 38, 44, 58